Advanced Structured Materials

Volume 39

Series editors

More information about this series at http://www.springer.com/series/8611

Moones Rahmandoust · Majid R. Ayatollahi

Characterization of Carbon Nanotube-Based Composites Under Consideration of Defects

Springer

Moones Rahmandoust
Griffith School of Engineering
Griffith University (Gold Coast Campus)
Southport
Queensland
Australia

and

Protein Research Center
Shahid Beheshti University, G.C.
Tehran
Iran

Majid R. Ayatollahi
Fatigue and Fracture Research Laboratory, Center of Excellence in Experimental Solid Mechanics and Dynamics, School of Mechanical Engineering
Iran University of Science and Technology
Tehran
Iran

ISSN 1869-8433 ISSN 1869-8441 (electronic)
Advanced Structured Materials
ISBN 978-3-319-00250-7 ISBN 978-3-319-00251-4 (eBook)
DOI 10.1007/978-3-319-00251-4

Library of Congress Control Number: 2015950020

Springer Cham Heidelberg New York Dordrecht London

Printed on acid-free paper

Springer International Publishing AG Switzerland is part of Springer Science+Business Media
(www.springer.com)

Preface

Composites are engineered materials that consist of two or more insoluble phases combined together: a continuous phase, known as the *matrix*, as well as some interdispersed components, known as the *reinforcing* elements. Now if at least one of the constituent phases of a composite material is less than 100 nm in size, this composite is commonly termed *nanocomposite*.

The matrix is typically the major constituent that provides durability for the overall composite and it can be for instance, a metallic, ceramic, or polymer material. The reinforcing inclusions are on the other hand, the structure's load carriers that can be in the form of fibers, particles, or flakes. This phase of the composite structure provides its stiffness and strength. Therefore, suitable selection of type, amount, and orientation of these components plays a very important role on the overall characteristics of the pertaining composite, such as its tensile and compressive strengths, fatigue strength and failure mechanisms, electrical and thermal conductivities, specific gravity, etc.

Composite materials mostly show more significant advantageous properties compared to monolithic materials. Monolithic metals and their alloys cannot always meet the demands of today's advanced technologies and performance requirements. It means that by incorporating reinforcements into, for instance, a metallic, ceramic, or polymer matrix, the properties of the matrix improves to a higher mechanical strength, more significant temperature stability, and better chemical durability. The existence of reinforcing elements, on the other hand, improves the structure's physical and chemical properties significantly.

Among all the variety of different fillers that can be used as a nanocomposite's reinforcing phase, *carbon nanotubes* (CNTs) have shown to be promising candidates for their very specific mechanical, electrical, optical, and thermal properties that will be explained later. Carbon nanotube-based nanocomposites, i.e., composite materials in which carbon nanotubes are used as the composite's reinforcing phase, are therefore very much interesting for scientists and scholars. There are many outstanding and remarkable applications that CNT-based composites can contribute to the world of science and industry. They are expected to influence many fields in terms of technology and industry. Either as stand-alone nanomaterials or as reinforcements in composites, they have applications in many diverse fields such as energy, signal processing, medicine, biotechnology, information technology, aerospace, agriculture, and environment. Therefore, exploring different properties of carbon nanotubes and carbon nanotube-based composite materials is of never-ending importance.

In the first three chapters of this volume, therefore, carbon nanotubes and carbon nanotube-based nanocomposites are introduced. Their material structure, major mechanical, electrical, and optical properties and their applications are further represented and the most common fabrication methods are also introduced. Finally, defects involved with carbon nanotubes, CNT-based composites are introduced, and the finite element techniques to model the introduced defects and to study their impact on the mechanical properties of the nanostructures are revealed for those who are interested in the area.

Chapter 4, then provides a detailed study on experimental aspects and challenges of fabricating carbon nanotube-based nanocomposites and controlling the appearance of defects in them. Finally, the major mechanical properties of the produced structures are reviewed and the effects of fabrication and processing steps on the resulted properties are introduced. Although mechanical properties of CNT-based composites are of major consideration in this volume, a specific subsection is devoted to highlight the special electrical conductance of the fabricated CNT composites.

The final chapter introduces CNT-fiber reinforced polymer composites specifically, as one of the most famous CNT-based composites and covers all major facts about the fabrication process of this nanocomposite type. The quality of the impact of modifying various fabrication factors of the polymer nanocomposites on its mechanical behavior is also explained with a detailed look at the tribological properties the composite.

Finally, we appreciate the moment to express our sincere gratitude to Professor Dr.-Ing. Andreas Öchsner, editor-in-chief of the Springer book series on "Advanced Structured Materials" for his helpful guidance. We would like to thank Dr. Shahin Shadlou, Dr. Mohsen Khalily, Mr. Iman Eslami Afrouz, Mr. Mehdi Mavalizadeh, Mr. Morteza Farsadi, Mr. Ali Ghavamian and Mr. Reza Barbaz for their valuable assistance. Dr. Mostafa Rahmandoust, Mrs. Azar Rezaie and Dr. Hamidreza Kazempour should also be appreciated here for their kind support. In conclusion, we are grateful to many scientists worldwide, for their novel contributions to the topic, and for devoting their lives to keep the flame of knowledge and science burning brightly and beautifully through human history. We also thank Springer publisher for recognizing the necessity of the topic and the importance of characterization of defects in nanostructured materials.

Moones Rahmandoust
Majid R. Ayatollahi

The original version of the book frontmatter was revised: For detailed information please see Erratum. The Erratum is available at 10.1007/978-3-319-18410-4_7

Contents

Synopsis

Composites are engineered materials that consist of two or more insoluble phases combined together, i.e., a continuous phase, known as the matrix, as well as interdispersed component known as the reinforcing phases. If at least one of the constituent phases of a composite material is less than 100 nm in size, e.g., the reinforcing phase, this composite is commonly termed a nanocomposite. Among all the variety of different fillers that can be used as a nanocomposite's reinforcing phase, carbon nanotubes have shown to be promising candidates for their very specific and remarkable mechanical and physical properties. Carbon nanotube-based nanocomposites, i.e., composite materials in which carbon nanotubes are used as the composite's reinforcing phase, are therefore very much interesting for many outstanding applications that they can contribute to the world of science and industry. This volume represents the characterization methods involved with carbon nanotubes and carbon nanotube-based composites, with a more detailed look on computational mechanics approaches, namely the finite element method. Special emphasis is given to studies that consider the influence of the existence of some imperfections in the structure of the nanomaterials on their mechanical properties. These defects may include random distribution of fibers in the composite structure, as well as atom vacancies, perturbation, and doping in the structure of individual carbon nanotubes.

Synopsis

List of Figures

List of Tables

Abbreviations

AFM	Atomic force microscope
BEM	Boundary element method
CFRP	Carbon fiber reinforced polymer
CNF	Carbon nanofiber
CNT	Carbon nanotube
DOS	Density of states
DWCNT	Double-walled carbon nanotube
ERM	Effective reinforcing modulus
FEM	Finite element method
FRP	Fiber-reinforced polymer
GFRP	Glass fiber-reinforced polymer
HDPE	High-density polyethylene
HRTEM	High-resolution transmission electron microscope
MD	Molecular dynamics
MWCNT	Multi-walled carbon nanotubes
Nano-Al	Nanocrystalline-aluminum
OAS	Optical absorption spectroscopy
PET	Poly ethylene terephthalate
PMC	Polymer-matrix composites
RM	Rule of mixture
RVE	Representative volume element
SPS	Spark plasma sintering
SWCNT	Single-walled carbon nanotube
TEM	Transmission electron microscope
UHMWPE	Ultrahigh molecular weight polyethylene

Symbols

$\vec{C}_h$	Chiral vector
θ	Chiral angle
a_0	Length of each unit vector
b	Carbon-carbon bond length
D, d	Diameter
t	Thickness
r	Distance
L	Length
A	Area
$\bar{m}$	Mass density
V_{LJ}	Lennard-Jones potential
F_{LJ}	Lennard-Jones force
σ	Stress
ε	Strain
F	Force
T	Torque
P_{cr}	Critical load
f	Resonance frequency
E	Young's modulus
G	Shear modulus
υ	Poisson's ratio
V_{CNT}	Volume fraction of CNTs
U	Strain energy
P	Pure axial load
M	Pure bending moment
T	Pure twisting moment
I	Second moment of area
J	Polar moment
Δb, $\Delta\alpha$ and $\Delta\beta$	Tensile, bending, and twist angle deformations

η_L	Length efficiency factor
η_o	Orientation efficiency factor
k_r, k_θ and k_φ	Molecular mechanics force constants
k	Spring constant
K	Effective length constant

Chapter 1
Introduction

1.1 Nanotechnology

Imagine how much control over resultant properties of a specific material you would have, if you could deposit each individual atom into a predefined arrangement toward a new material. This "bottom-up" materials design is becoming a reality by means of nanotechnology. Hence, with this technology, one is able to fabricate and manipulate materials that comprise of a small number of atoms, with features on the nanometer scale (10^{-9} m), i.e. one-billionth of a meter, or in other word, materials with dimensions approximately 1,000 times smaller than the diameter of a human hair! To put this into perspective, add this fact that it is now even possible to push individual atoms around a surface using specialized techniques.

Physics Nobel Prize winner, Richard Feynman, gave the first lecture regarding the prospective applications for nanoscale materials. His talk, entitled "There's Plenty of Room at the Bottom," was delivered on 29 December 1959, at the annual American Physical Society meeting, which contained references to a future world that was never before imagined, a world that is built by molecular machines with atomic accuracy. Feynman pointed out that designing material atom-by-atom is a real possibility, as it would not violate any physical laws (Fahlman 2011). This lecture is often considered as the inspiration of today's nanotechnology.

The term "nanotechnology", on the other hand, was first used by Japanese scientist, Norio Taniguchi from the Tokyo University of Science, in 1974, at the International Conference on Precision Engineering (ICoPE). He used this term to refer to "a production technology that has extra high accuracy and ultra fine dimensions, i.e., the preciseness and fineness on the order of 1 nm in length (Taniguchi 1974)."

However, the real nanomaterials revolution was after the discovery of fullerenes in 1985. A fullerene, shown in Fig. 1.1, that is totally composed of carbon atoms,

M. Rahmandoust and M.R. Ayatollahi, *Characterization of Carbon Nanotube-Based Composites Under Consideration of Defects*, Advanced Structured Materials 39,
DOI 10.1007/978-3-319-00251-4_1

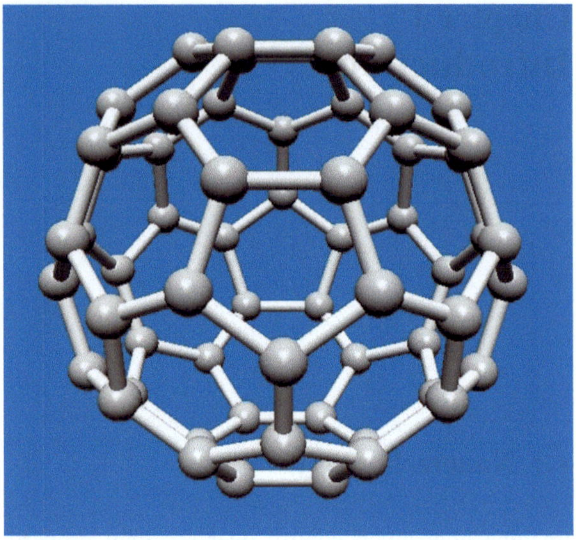

Fig. 1.1 C60 carbon fullerene

was discovered by Richard Smalley, Robert Curl, James Heath, Sean O'Brien, and Harold Kroto at Rice University (Edwards 2006). This discovery, followed by the discovery of carbon nanotubes in 1991 (Iijima 1991) produced a lot of excitement in terms of possible applications of these carbon-based nanomaterials, for their lightweight and high-strength.

1.2 Nanostructured Materials

From nanomaterials point of view, NASA suggests the following thorough description for the term nanotechnology:

> The technology of the creation of functional materials, devices and systems through control of matter on the nanometer length scale (1 to 100 nm), and exploitation of novel phenomena and properties (physical, chemical, biological) at that length scale (NASA 2010).

Generally, nanostructures are materials with at least one dimension in nanometer scale, i.e. <100 nm. These materials can be categorized into four main groups in terms of their macroscopic dimensions, namely zero-, one-, two- and three-dimensional nanostructures. Zero-dimensional nanostructures (0D) are those that have all dimensions restricted in the nanometer size and hence, have no dimensions in macroscale. Nanoparticles, quantum dots, and nanodots and in general isolated nanoclustures are categorized among this group. Same way, one-dimensional nanostructures (1D) are those with one dimension in macroscale and two in nanoscale. Nanofibers, nanowires, nanorods, and of course nanotubes that are specifically important in our scope are the members of this group. After coatings, thin-films and multilayers are considered two-dimensional (2D) as a result

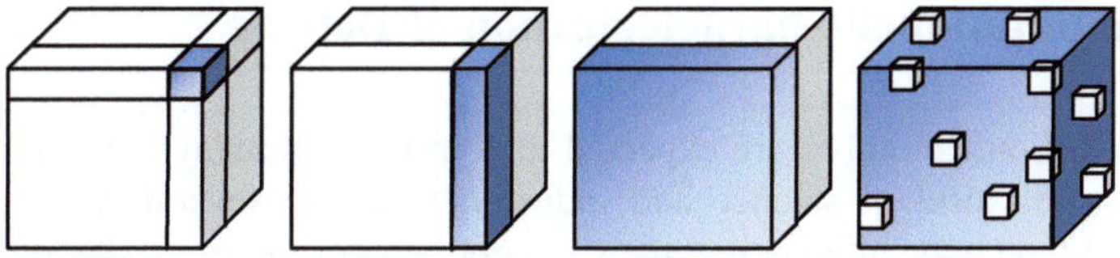

Fig. 1.2 Nanostructures categorized base on macroscopic dimensions

of being confined in only one nanoscale dimension, three-dimensional (3D) structures should be introduced with all of their dimensions in macroscale. Then one may wonder why and how a macroscale structure could be categorized as a nanostructure; the answer lies inside structure!

Any type of 0D, 1D or 2D nanomaterials distributed within a macroscopic liquid or solid medium will create a 3D nanomaterial (Gusev and Rempel 2004). Nanocomposites for instance are categorized as three-dimensional nanostructure, comprising of at least two or more insoluble phases from which at least one phase is <100 nm in size. Considering that in composites, there is in general a continuous phase, known as the matrix, as well as some interdispersed components, known as the reinforcing elements or fibers, when the so-called fibers are in nanoscale size, the composite is commonly termed as a *nanocomposite*. Figure 1.2 shows zero- to three-dimensional nanostructures schematically, abbreviated as 0D to 3D, respectively.

1.3 Nanomaterials' Emerging Applications

Although many exciting applications were imagined for carbon-based nanomaterials, the achievements in this respect have not been still remarkable. Nanotubes in particular have been more successful in bringing about a revolution in materials technology. Bundles of billions of carbon nanotubes can approach any imagined limit of tensile strength. The intrinsic strength of carbon nanotubes is theoretically the highest and use of bundles will help bypassing any possible breakdowns that may occur due to propagation of microscopic defects. Estimates about the potential tensile strength of carbon nanotube based composites vary, but it is expected that a 1 m rod of carbon nanotubes may be 50–100 times stronger than steel, with only one-sixth of its weight. Imagine how such a material could influence mechanical engineering, civil and building construction, aircraft, and automobile technologies (Encyclopædia-Britanica 2014).

However, the technology of producing long and perfectly ordered bundles of billions of nanotubes is not yet achieved but the more realistic approach of using carbon nanotube-based composite materials with improved properties is reached for many possible applications that are even feasible for biomedical purposes such as micro- or nano-surgeries and drug delivery as well.

1.4 Computational Characterization of Defects

Although carbon nanotubes are expected to contribute in many outstanding applications in the world of science and industry, there are several types of imperfections and defects influencing their exceptional properties. The defects that occur during the process of manufacturing may affect their potential function as a part of a nanodevice. Defects are an inevitable part of all devices. Although in the macro world, the existence of defect does not influence the functionality of devices extensively, in the world of nano, even minor defects can alter the basic properties of the device significantly. Hence, the study of the defects is a major topic that should be investigated very carefully in nanoscience and technology (Rahmandoust and Öchsner 2015). These defects may be majorly categorized into two main groups, i.e. topological defects occurring in the structure of carbon nanotubes, compared to the ideal arrangement of atoms, as well as defects that arise due to the doping of foreign atoms to the structure of carbon nanotubes. Introduction of any defects to the ideal structure of carbon nanotubes will result in a change in physical properties. These changes are in many cases desirable, namely for producing specific enhanced conductivity or semiconductivity or for boosting the sensing ability of the tube, where a controlled introduction of defects is required. On the other hand, the consequences of the existing defects may have undesirable influence on the properties of carbon nanotubes as well, namely decrease of mechanical strength of the structure, or lowered conductivity, that should be considered majorly by in-advance computational characterizations, before the nanostructure is employed as a functioning part of a device.

References

Edwards, S.A.: The Nanotech pioneers: where are they taking us?. Wiley-VCH, Weinheim (2006)

Encyclopædia-Britanica: Fullerene (2014). http://www.britannica.com/EBchecked/topic/221916/fullerene

Fahlman, B.D.: Materials chemistry. Springer, Heidelberg (2011)

Gusev, A.I., Rempel, A.A.: Nanocrystalline Materials. Cambridge International Science Publishing, Cambridge (2004)

Iijima, S.: Helical microtubules of graphitic carbon. Nature **354**, 56–58 (1991)

NASA: NASA Website (2010). http://www.ipt.arc.nasa.gov/nanotechnology.html

Rahmandoust, M., Öchsner, A.: Defects involved with carbon nanotubes. In: Kharisov, I.B. (ed.) CRC Concise Encyclopedia of Nanotechnology. CRC Press. p.in press, Kentucky (2015)

Taniguchi, N.: On the basic concept of nanotechnology. In: International Conference on Precision Engineering, Tokoyo (1974)

Chapter 2
Carbon Nanotubes

2.1 Introduction

Carbon nanotubes have attracted significant attention since their discovery, as unique valuable nanostructures, with many outstanding properties, leading them toward being applied in a wide variety of novel and amazing applications. The extraordinary properties of carbon nanotubes stem mostly from their perfect hexagonal structure, as well as their high length to diameter or aspect ratio, which is specific to most of nanostructured materials. In other terms, nanomaterials are all uniquely beneficial in many of today's emerging applications, only as a result of their high aspect ratio. This specific characteristic make them interact with their surrounding environment more efficiently, especially when it comes to adsorption properties and interaction in a gaseous environment. In this chapter, the structure of carbon nanotubes and the methods of fabricating them will be introduced and the defects associated with their structure will be explained in detail.

2.2 Carbon and Carbon-Based Nanomaterials

Carbon, the fourteenth element of the periodic table, with symbol C, is a unique element. This exceptional element tends to bond with various atoms differently and hence, presents a variety of exceptional physical, chemical and biological properties. The atomic number of carbon is six and therefore, it has four electrons in its outer shell. These four electrons, known as valance electrons, are available for participating in different types of chemical bonding. Hence, depending on its covalent bonding type, different allotropes of carbon are formed that exhibit tremendously different properties. For instance, graphite and diamond are both allotropes of carbon, though graphite is soft, black and a very good electrical

M. Rahmandoust and M.R. Ayatollahi, *Characterization of Carbon Nanotube-Based Composites Under Consideration of Defects*, Advanced Structured Materials 39,
DOI 10.1007/978-3-319-00251-4_2

conductor, but diamond is very hard, transparent and a very bad conductor! Carbon nanotubes and fullerenes, on the other hand, are the nanoscaled allotropes of carbon with many outstanding properties. Figure 2.1 shows these different allotropes of carbon.

As explained before, the discovery of carbon buckyballs or fullerenes in 1985 was one of the initial stages of the development of nanotechnology. The path of progress was further developed in 1991, when carbon nanotubes were first observed and their remarkable properties were introduced to the world of science and technology. Carbonous nanomaterials had shown to be promising candidates as outstanding engineering materials for their exceptional mechanical, thermal, optical and electrical properties. Having four valance electrons, carbon can get involved in chemical bonding in three different orbital hybridizations, which are the *sp*, sp^2 and sp^3 hybridizations, happening for instance in alkynes, graphite and diamond, respectively. The hybridization of the atomic orbital of carbons

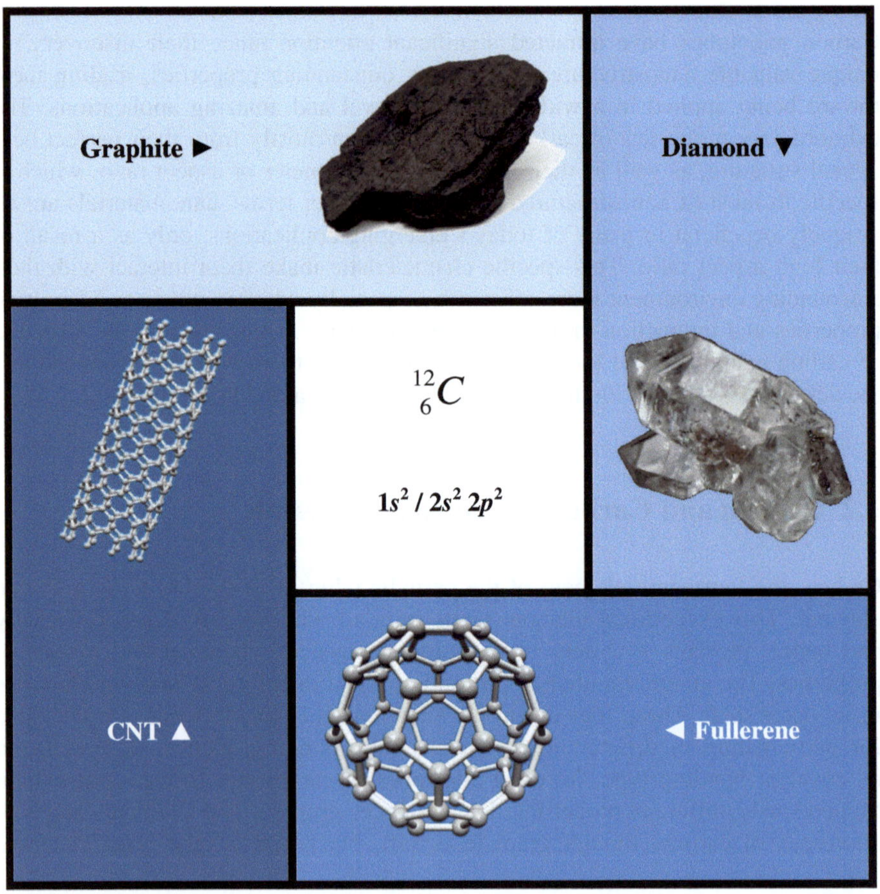

Fig. 2.1 Carbon atom and its allotropes

in carbon nanotubes and fullerenes is of type sp^2, i.e. similar to that of graphite, making a perfect hexagonal array of atoms (Rahmandoust and Öchsner 2012).

In sp^2 hybridization, three of the valance shell atoms are involved in chemical bonding and hence, there is one free electron, called the *π-electron*. Materials with extended π-electron clouds, like carbon nanotubes, are called the π-electron materials that exhibit many exceptional thermal and electrical conductivity properties (Ahmadi et al. 2012). The sp^2 C–C bond of CNTs is considered one of the strongests in solid materials; thereby CNTs are expected to yield exceptionally high stiff and strong mechanical properties (Dresselhaus et al. 1995).

2.3 Atomic Structure of Carbon Nanotubes

The hexagonal arrangement of atoms in carbon nanotubes and graphene sheets can be described in terms of the tube's chirality or helicity, using the chiral vector, $\vec{C}_h$ or the chiral angle θ. The chiral vector, $\vec{C}_h$, also known as the roll-up vector, is defined by the following equation:

$$\vec{C}_h = m\vec{a}_1 + n\vec{a}_2, \tag{2.1}$$

where the integers n and m are the number of steps along the unit vectors $\vec{a}_1$ and $\vec{a}_2$, as defined in Fig. 2.2. The chiral angle θ determines the amount of *twist* of the carbon nanotube.

Hence, carbon nanotubes are fundamentally classified into three main categories according to their chiral vector, $\vec{C}_h$. A *chiral* (m,n) nanotube in general has unequal, non-zero m and n integers, i.e. $m \neq n \neq 0$ and the chiral angle θ is between its two limits ($0° < \theta < 30°$), whereas for the two limiting cases,

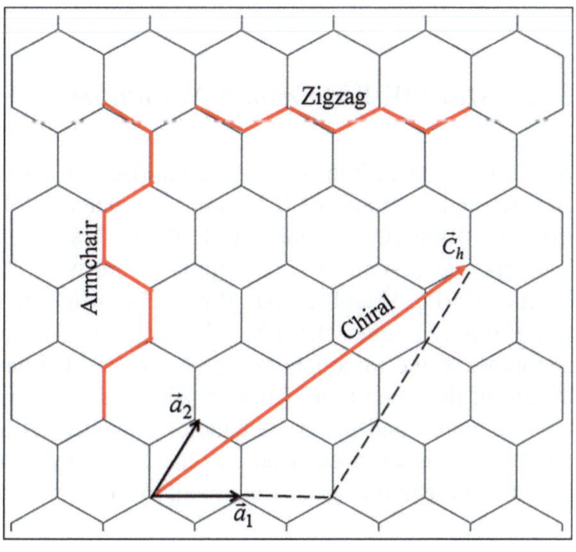

Fig. 2.2 Schematic view of a single-walled zigzag, armchair and chiral carbon nanotubes

i.e. *zigzag* nanotube with chiral angle equal to zero ($\theta = 0°$) and *armchair* nanotube with ($\theta = 30°$), we have $m \neq n$, $n = 0$ and $m = n \neq 0$, respectively. These last two limiting cases are generally known as symmetric carbon nanotubes (Dresselhaus et al. 1995). The chiral, armchair and zigzag carbon nanotubes sometimes exhibit different properties. For instance, chiral, and zigzag nanotubes show semiconducting behavior, whereas armchair carbon nanotubes are categorized as metal and are considered conducting materials (Ahmadi et al. 2012). In addition to chirality, the number of layers is also another factor that is considered for categorizing carbon nanotubes, hence they are categorized as *single-walled carbon nanotube* (SWCNT) and *multi-walled carbon nanotube* (MWCNT) which are explained in detail below.

2.3.1 Single Walled Carbon Nanotubes

A single-walled carbon nanotube can be imagined as a graphene sheet that has been rolled into a tube or cylinder with a diameter in range of 0.4–3 nm. The thickness of the tube's wall is considered equal to that of a graphene sheet, which is about 0.34 nm (To 2006; Tserpes and Papanikos 2005; Li and Chou 2003). For a given (*m*,*n*) carbon nanotube, the radius of the tube can be calculated using chiral vector integers, *m* and *n*, as follows:

$$R_{\mathrm{CNT}} = \text{Length of } \vec{C}_h/2\pi = \frac{a_o\sqrt{m^2+mn+n^2}}{2\pi}, \tag{2.2}$$

where a_0 is the length of each unit vector. In carbon nanotubes, the sp^2 carbon–carbon bond length is $b = 0.142$ nm and therefore a_0 is equal to $\sqrt{3}b$ (Dresselhaus et al. 1995; Natsuki et al. 2004).

2.3.2 Multi Walled Carbon Nanotubes

A multi-walled carbon nanotube comprises of two to 50 coaxial nanotubes with an inter-layer spacing of 0.34 nm. The diameter of MWCNTs generally ranges from 3–30 nm (Dresselhaus et al. 1995). These nanostructures were discovered and reported first in 1991 by Iijima at the NEC Laboratory in Tsukuba, Japan (Iijima 1991). He observed MWCNTs, using a *high-resolution transmission electron microscope* (HRTEM), 2 years before discovering single-walled carbon nanotubes at the same laboratory (Iijima and Ichihashi 1993). However, approximately at the same time, Russian scientists also reported that they observed carbon nanotubes and nanotube bundles (Gal'Pern et al. 1992; Kosakovskaya et al. 1992). Their discovered structures had generally a much smaller length to diameter ratio compared to what was found in Iijima's laboratory. Bethune also reported the experimental discovery of single-walled carbon nanotubes in 1993, at the IBM Almaden Laboratory in California, the Unites States (Bethune et al. 1993).

As explained before, hybridization of carbon atomic orbital in carbon nanotubes is of type sp^2, in which each atom is joined to three other neighbors in a planar trigonal arrangement forming a hexagonal sheet of covalence bonds. In case of MWCNTs, in addition to the above mentioned strong covalence chemical interactions, individual sheets or cylinders interact with each other by non-covalent interactions that can be adequately described by weak van der Waals force, using the Lennard-Jones potential (Kalamkarov et al. 2006), presented in Eq. (2.3).

$$V_{\mathrm{LJ}} = 4\varepsilon\left[\left(\frac{\sigma}{r}\right)^{12} - \left(\frac{\sigma}{r}\right)^{6}\right] \tag{2.3}$$

In Eq. (2.3), the terms σ (in nm) and ε (in kJ/mol) are the *Lennard-Jones parameters*. These parameters are material-specific and determine the nature and strength of the interaction. For a carbon–carbon non-covalent interaction, the values of the Lennard-Jones parameters are $\sigma = 0.3851$ nm and $\varepsilon = 0.4396$ kJ/mol. The diagram of the Lennard-Jones potential interacting between two carbon atoms is presented in Fig. 2.3.

As it is shown in the diagram, σ is by definition, the distance at which the potential between the two interacting particles becomes zero and ε is the minimum potential acting on them. As a result of the Lennard-Jones potential, the Lennard-Jones force acts on the particles. The Lennard-Jones force is strongly repulsive when the particles approach too close to each other and is mildly attractive when they are far away. The Lennard-Jones force, given below, is in general considered as a weak force. The critical distance, shown in Fig. 2.4, is the distance at which the force becomes zero is named as r_o and is equal to $r_o = \sqrt[6]{2} \cdot \sigma$

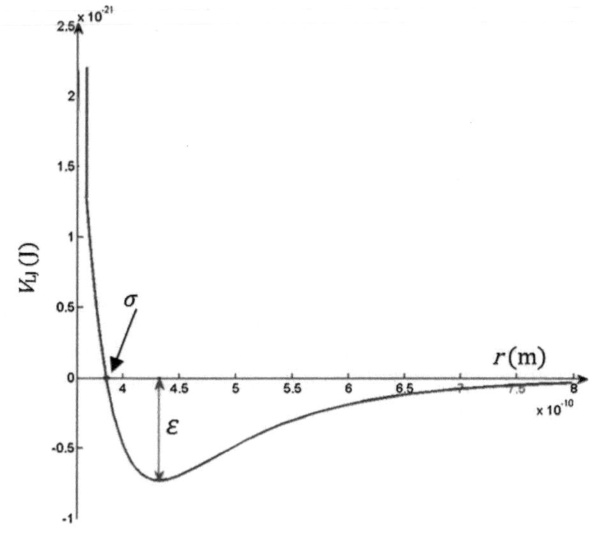

Fig. 2.3 Carbon–carbon Lennard-Jones interaction

Fig. 2.4 The Lennard-Jones force

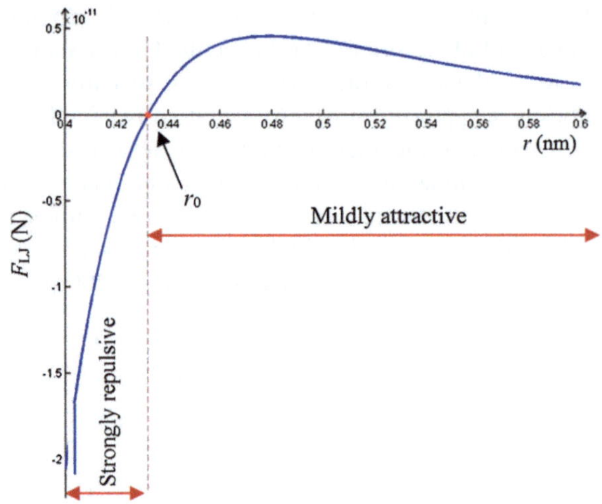

$$F_{LJ} = \frac{dV_{LJ}}{dr} = \frac{4\varepsilon}{r}\left[-12\left(\frac{\sigma}{r}\right)^{12} + 6\left(\frac{\sigma}{r}\right)^{6}\right] \tag{2.4}$$

Although introducing the Lennard-Jones interaction leads to more accurate results with respect to the analytical solution, in modeling multi walled carbon nanotubes, investigations show that obtained results, when this weak interaction was not defined in the model are reliable and the difference is ignorable, especially when longer tubes are modeled and studied (Rahmandoust and Öchsner 2012a, b). However, in many investigations, this type of weak bonding between the layers of carbon nanotubes is defined to achieve more accurate results.

Other than the Lennard-Jones potential, there is another common model for showing two-body interactions, known as the Morse potential function, introduced in the following equations:

$$V(r) = \varepsilon[e^{2\beta(\rho-r)} - 2e^{\beta(\rho-r)}], \tag{2.5}$$

$$F(r) = 2\varepsilon\beta\left[e^{2\beta(\rho-r)} - 2e^{\beta(\rho-r)}\right], \tag{2.6}$$

where β is an inverse length-scaling factor, ρ and ε are the equilibrium bond length and displacement energy, respectively. The Morse potential function has been modified for various types of materials and interfaces. In chemistry, physics and engineering, the Lennard-Jones and Morse potentials are commonly used in molecular dynamics simulations, based on the two-body approximation (Ghadyani and Rahmandoust 2015).

2.4 Properties of Carbon Nanotubes

As mentioned earlier, the discovery of carbon nanotubes produced a lot of excitement among scientists, as a result of their remarkable properties and the potential extraordinary application that they could introduce to the world of science and technology. The mechanical strength per volume of carbon nanotubes is much greater than conventional materials. Furthermore, their outstanding electrical properties have made them a suitable option in fabricating flat panel TV and monitor displays, batteries and many other electronic devices. Scientists were correctly dreaming about expanding the borders of their imagination on the new and emerging capabilities that these materials could introduce as a part of many small and smart devices.

2.4.1 Mechanical Properties

It is widely recognized that the elastic modulus of a material is closely related to the chemical bonding of its constituent atoms. The sp^2 carbon–carbon bonding, observed in carbon nanotubes, is considered one of the strongest solid materials; thus, CNTs are expected to yield exceptionally high stiff and strong mechanical properties (Sie 2009). Using different study techniques, Young's modulus of carbon nanotubes, obtained from experimental and numerical investigations, has been reported to be very high, i.e. about 1 TPa.

2.4.1.1 Elastic Modulus

Numerous methods have been employed to study various mechanical properties of CNTs in their perfect and defective forms. Lu posits values of the Young's modulus of CNTs in a range between 0.97 to 1.11 TPa, using empirical force-constant model (Lu 1997), while Krishnan et al. measured Young's modulus for 27 isolated single-walled carbon nanotubes experimentally using transmission electron microscope (TEM) and achieved values between 0.9 to 1.7 TPa, with a mean of 1.25–0.35/+0.45 TPa (Krishnan et al. 1998). Tight-binding and ab initio quantum mechanical approaches were also applied to investigate Young's modulus of carbon nanotubes in 1999 by Hernández et al. and Sánchez-Portal et al. respectively. Hernández reported a value between 1.22 to 1.25 TPa for different types of CNT chiralities investigated (Hernández et al. 1999a), while Sánchez posits Young's modulus of carbon nanotubes to be almost equal to that of graphene, regardless of the tube's chirality and type (Sánchez-Portal et al. 1999). More recent experimental investigations, using atomic force microscope (AFM), resulted in obtaining Young's modulus value = 1.2 TPa (Tombler et al. 2000).

Li and Chou obtained values from 0.89 to 1.033 TPa, using continuum structural mechanics (Li and Chou 2003), whereas Natsuki obtained the elastic modulus of CNTs in the range 0.5 TPa to more than 1.1 TPa, depending on the wall thickness, using the same approach (Natsuki et al. 2004). To reported Young's modulus about 1.024 TPa based on continuum mechanics studies (To 2006). Molecular dynamics (MD) was also applied extensively to study the mechanical properties of carbon nanotubes. Chang and Gao used the technique to predict a value of the Young's modulus of 1.06 TPa (Chang and Gao 2003). WenXing et al. considered different chiralities and tube lengths and obtained elastic modulus ranging between 0.904 to 0.939 TPa (WenXing et al. 2004). Molecular mechanics is another very common scientific approach used to investigate the elastic properties of CNTs. In an investigation with the purpose of obtaining the fracture behavior of carbon nanotubes, Duan et al. used this technique to report values between 1.05 and 1.34 TPa for Young's modulus (Duan et al. 2007). Fang et al. observed the influence of different tube diameters on elastic properties of CNTs that ranged from 0.95 to 0.99 TPa for (Fang et al. 2007). First-principles approach leaded to values around 0.918 TPa for zigzag and about 0.874 TPa for armchair CNTs (Song et al. 2010), whereas by the finite element method (FEM) investigations, Young's modulus of single- and multi-walled CNTs was obtained in the range 1.03–1.05 TPa (Rahmandoust and Öchsner 2012a, b).

Other than single-walled carbon nanotubes, multi-walled carbon nanotubes have also been studied by different scientists. Kalamkarov used FEM to analyze MWCNT and studied their mechanical properties. Young's modulus of the MWCNT, obtained from this research, ranges from 1.32 to 1.58 TPa (Kalamkarov et al. 2006), whereas in a more recent investigation Fan reported Young's modulus to be ranging from 0.99 to 1.03 TPa using the finite element method (Fan et al. 2009). Depending on the approach and the energy potential used, Poisson's ratio of the carbon nanotube was obtained around 0.05–0.28 (Natsuki et al. 2004; Tserpes and Papanikos 2005; Sie 2009; Rahmandoust and Öchsner 2009).

2.4.1.2 Shear Modulus

Shear modulus of carbon nanotubes has also been evaluated by two types of tests, i.e. torsion and tensile tests for both SWCNTs and MWCNTs. The results are generally reported to be in the range of 0.2 and 0.5 TPa, which is considered a high shear modulus. However, the vast reported range is due to the applied method of obtaining the results. According to a recent investigation, the remarkable difference observed between the results from torsion and tensile tests, is due the non-isotropic behavior of carbon nanotubes (Ghavamian et al. 2013).

Jin and Yuan evaluated the mechanical properties of single-walled carbon nanotubes by numerical simulation, using both energy and force approaches. They predicted the shear modulus about 0.546 and 0.492 TPa, respectively (Jin and Yuan

2003). Li and Chou acquired shear modulus of single-walled carbon nanotubes in the range of 0.22–0.48 TPa, by using a structural mechanics approach (Li and Chou 2003). In the case of MWCNTs, Li and Chou obtained the shear modulus value about 0.40 $\pm$ 0.05 TPa, which was slightly lower than the results, obtained for SWCNTs, while Yu applied a molecular dynamics method in which single-walled carbon nanotubes were simulated under the torsion test to evaluate their mechanical properties. According to their results, the value of shear modulus changed between 0.37 and 0.5 TPa (Yu et al. 2004). Furthermore, Natsuki developed an analytical method for modeling the elastic properties of single-walled carbon nanotubes. They obtained the shear modulus in the range of 0.32 and 0.37 TPa (Natsuki et al. 2004).

Tserpes and Papanikos obtained the shear modulus of different single-walled carbon nanotubes, varying from 0.283 to 0.504 TPa by defining three-dimensional finite element models (Tserpes and Papanikos 2005). Kalamkarov team used two different continuum-based approaches to model the behavior of CNTs and investigated their mechanical properties. The first approach which was an analytical one, resulted in the shear modulus equal to 0.32 TPa for single-walled CNT and the second approach which was a numerical method, using the finite element method, predicted shear modulus between 0.14 and 0.47 TPa for single-walled and 0.44–0.47 TPa for three- and four-walled carbon nanotubes (Kalamkarov et al. 2006). In addition, To also employed finite element method in which the Poisson's ratio effect was included in the determination of mechanical properties of SWCNTs. This investigation yielded the shear modulus equal to 0.47 TPa (To 2006). A combination of molecular mechanics and continuum mechanics was applied to evaluate the properties of CNTs by Wu. The value of shear modulus obtained by this method was in the range of 0.32–0.41 TPa for SWCNTs (Wu et al. 2006).

The shear modulus of CNTs was also derived by putting the nanostructure under tensile test. Fan and his research team applied the FEM simulation technique to investigate the mechanical properties of MWCNTs. According to their results, the shear modulus of CNTs was calculated in the range of 0.35–0.45 TPa for single-, double- and triple-walled carbon nanotubes (Fan et al. 2009). Rahmandoust and Öchsner applied the same technique based on the substitution of covalent bonds between atoms by beam elements to model multi-walled carbon nanotubes, and by defining non-material links between the layers to introduce the van der Waals force. They finally acquired the value of shear modulus of the multi-walled carbon nanotubes from 0.37 to 0.47 TPa (Rahmandoust and Öchsner 2012a, b), whereas for SWCNTs they reported 0.48 TPa as shear modulus (Rahmandoust and Öchsner 2009). Ghavamian used finite element method to compare the shear modulus obtained from torsion or tensile tests for single and multi walled carbon nanotubes. According to his investigations, increasing the number of CNT walls, leads to a decrease in the value of the shear modulus (Ghavamian et al. 2013).

2.4.1.3 Buckling Behavior and Resonant Frequency

Buckling behavior and natural resonance frequency of single- and multi-walled carbon nanotubes were also investigated by many scholars. Carbon nanotubes are very vulnerable to buckling, as they are long and hollow structures. However, the results show that they are at the same time capable of retaining their buckling capacity and recover their elasticity (Rahmandoust and Öchsner 2012a, b; Ghavamian and Öchsner 2012; Wang et al. 2010; Fan et al. 2009). A brief review on the investigations about the buckling properties of CNTs has been published in 2010 (Wang et al. 2010). Besides, investigations show that adding small portion of CNT's to neat resin, can improve the fundamental frequencies of CNT reinforced composite remarkably, without imposing any striking change in mass density of material (Fereidoon et al. 2013; Formica et al. 2010). Hence, vibrational behavior of CNTs had been investigated accurately as one of the fundamental characteristics of intact/perfect CNT's (Moghadam et al. 2014; Ghavamian and Öchsner 2013a, b). On the other hand, as a result of their small size, carbon nanotubes can make acceptable resonators in signal processing systems as well. The resonant frequency enhances as the size of the resonator decreases and higher resonance frequency is equal to higher sensibility (Santos 1999).

2.4.2 *Electrical Properties*

The electrical properties of carbon nanotubes were in fact among those first interesting points found about these spectacular structures by scientists. Compared to conventional silicon-based electronics, carbon based nanoelectronics promise greater flexibility and advantages conferred by miniaturization (McEuen 1998). Carbon nanotubes are also considered as potential candidates for molecular-electronics (Vouris 2002) and many remarkable properties of them, peculiarly those that are due to their low dimensionality can be harnessed and gainfully employed in this field (Kelly 1995). However, large scale assembly of CNTs, at a level that would be striking to system designers is still challenging in many applications such as in devices where power savings, radiation hardness, and reduced heat dissipation are in major considerations (DeHon 2002).

In this section, firstly, the classification of CNTs into conductors and semiconductors is discussed and then, the electrical properties of single- and multi-walled CNTs are explained to include significant features of electrical transport, which can affect interpretation of measurements and suitability of them being used in advanced electronics.

2.4.2.1 Conducting and Semiconducting CNT

As explained before, carbon nanotubes can be categorized to three various main types according to their helicity, namely zigzag, armchair and chiral tubes. These different types are described using two integers n and m. According to this method of categorizing, any nanotube can be described by (m,n) for chiral nanotube, $(m,0)$ for zigzag and (m,m) for armchair carbon nanotube, creating variety of diameters and twisting angles. From the electronic properties point of view, carbon nanotubes can behave both as semiconducting and metallic, depending on their twisting angel. In other words, the helicity of the nanotube will influence the band gap of its density of states (DOS), varying from 0 eV to a range between 20 and 2 eV, and hence producing both metallic and semiconducting nanotubes out of the same material, respectively (Ahmadi et al. 2012). Many experimental and theoretical studies have focused on finding the relationship between the CNT's atomic structures and its electronic properties, as well as its electron-electron and electron-phonon interaction effects (Dekker 1999; Tanaka et al. 1997).

According to the available investigations, a given nanotube (m,n) nanotube exhibits metallic conducting properties with zero eV bandgap, if $2m + n = 3N$, where N is an integer equal to 0, 1, 2, etc. and therefore, they have electrons in their conduction bands at room temperature, conduct electricity very well, and are called metallic nanotubes. On the other hand, the tubes are semiconductor if $2m + n \neq 3N$, with a band gap varying inversely with the diameter of the tube. Regardless of the fact that it is necessary to refine the above condition by consideration of bond alternation, experiments on the electrical-conductivity measurement and studies on the other related solid-state properties of CNTs confirm that this rather simple but impressive prediction is capable of giving realistic analysis of the structure's electrical behavior (Ahmadi et al. 2012). Thus, (m,m) armchair and $(3m,0)$ zigzag carbon nanotubes are all metallic and the rest of zigzag CNTs and chiral tubes are semiconducting unless they satisfy the introduced $2m + n = 3N$ rule. In multi-walled Carbon nanotubes, however, modification of the electronic structure, especially in the metallic state, shows that due to symmetry reasons, the electronic structure of the inner tube does not seriously affect the metallic property of the outer-most (Saito et al. 1998).

In an individual carbon nanotube, there are many electrons at different energy levels within the tube's structure. As Fig. 2.5 illustrates schematically, the relative abundance of electrons as a function of the electron energy, which is much higher in the shaded layer, that is the valance band, while the unshaded area includes the energy levels that could be populated by electrons in carbon nanotubes. A photon

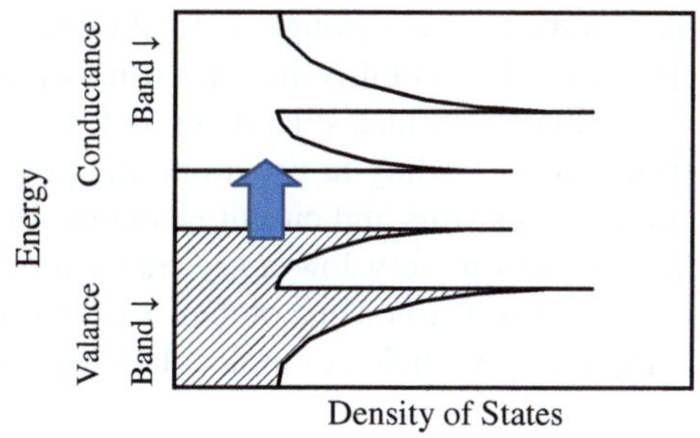

Fig. 2.5 Relative abundance of electrons as a function of energy level

whose energy is equal to the difference between a filled and an unfilled state can be easily absorbed, causing an electron transition from the shaded valance band energy state to the unshaded conductance band state. Having such photon absorptions spectrum characteristics, in the ultraviolet to near-infrared wavelength range, one can identify the (m,n) nanotube type by in a sample by analysis of their light absorption spectrum.

Charge carriers such as electrons and holes can move very easily through carbon nanotubes but when dealing with bundles (ropes) of carbon nanotubes, electrons will not only flow down individual tubes, but occasionally they can jump from one tube to another as well. Therefore, carbon nanotube ropes have been known as one of the most electrically conductive fibers ever (Thess et al. 1996). As the electrons within a nanotube are highly polarized, the energy levels of carbon nanotubes depend very much on their environment as well. For instance, application of an external field can shift the relative energies of these states remarkably towards enhancing or lowering their conductivity, making them candidates for printable semiconductor in transistor applications. Likewise, the presence of some types of foreign atoms or impurities can shift these energy levels and affect conductivity and optical behavior of CNTs significantly. In this way, custom functionalization of nanotubes can lead to applications of the nanomaterial as sensors or as probes for detecting cell abnormalities in biomedicine.

2.4.2.2 Ballistic Transport in CNTs

Furthermore, the nanostructure is considered a one-dimensional material and hence it is capable of exhibiting ballistic or quasi-ballistic transport along its main axis (Tans et al. 1998; Wei et al. 2001; Javey et al. 2003; Dürkop et al. 2004). Individual carbon nanotubes have been observed to conduct electrons ballistically, that is, with no scattering, for distances of a few microns (Tans et al. 1998). Ballistic conduction permits substantial current flow with minimal heat generation, which allows nanotubes to carry enormous amounts of current. In fact, carbon nanotubes can carry the highest current density of any known material, with conductivities measured as high as 10^9 A/cm^2 (Wei et al. 2001), whereas copper fails at current densities of about 10^7 A/cm^2, so nanotubes can carry current densities about 100 times that of traditional metal conductors. Another advantage of ballistic conduction is that the charge carriers can move very rapidly through nanotubes under the influence of an electric field. The charge carrier mobility in carbon nanotubes can exceed that in traditional electronic materials like silicon by at least an order of magnitude higher (Dürkop et al. 2004), leading to providing the possibility of manufacturing ultra-high-frequency transistors and circuit elements. Finally, as another consequence of ballistic conduction, very low-noise transistors can be fabricated by nanotubes (Javey et al. 2003). Carbon nanotubes are therefore expected to play a central role in many emerging fields of molecular electronics.

2.4.3 Optical Properties

In the late 1990s, theoretical and experimental studies reported the remarkable optical properties of carbon nanotubes after Margulis reported the first theoretical study on the very high third order nonlinearity of CNT and Kataura et al. reported the nonlinear optical properties of single-wall CNT for the first time (Kataura et al. 1999; Margulis 1999). Since then, various potential applications of CNTs have proposed and attracted significant attention in photonics research community. CNTs exhibit an exceptionally high third-order optical nonlinearity and nonlinear saturable absorption with ultrafast recovery time and broad bandwidth operation (Martinez and Yamashita 2011). Thus, they are becoming key components towards the development of fiber lasers, nonlinear photonic devices and several emerging next generation devices, both from an academic and industrial point of view.

2.4.3.1 Saturable Absorption

The optical absorption of CNTs is of saturable, intensity-dependent nature, it is, therefore, a suitable material to employ for specific applications like passively mode-locked laser operation. Saturable absorption is a result of the excitonic absorption in semiconductor nanotubes (Saito et al. 1992). The density of states of a semiconducting CNT can be numerically calculated. In fact, the electronic density of states of CNT, originates from the slicing of the graphene density of states, due to the quasi-1D confinement of the electrons, which leads to the presence of the characteristic Van Hove singularities in their electronic DOS. Semiconducting CNTs absorb photons in the same way an ordinary direct bandgap semiconductor would do when the energy of the photon is equal to the bandgap of the CNT. The CNT's bandgap is inversely proportional to the diameter of the tube. It is particularly fortunate that the diameters of CNTs typically ranges from 1 to 2 nm, which leads to an optical absorption that overlaps the 1550 nm window commonly used for fiber optic technologies.

Current processes for bulk fabrication of CNTs do not provide complete control over the chirality and diameter of the fabricated CNTs and produces a combination of metal and semiconductor SWCNTs and MWCNTs. However, separating the fabricated CNTs according to their chirality is possible through post-fabrication processes. In actual fact, depending on the expected application, a wide or a very narrow distribution of CNT diameters may be beneficial. Hence, the ability to control the mean diameter and diameter distribution in the fabrication, purification and separation processes is very helpful in expanding the application chances of CNTs in photonics science and industry. On the other hand, in terms of absorption for instance, for ultra-short pulse generation, a saturable absorber with a fast recovery time is required, while a slower recovery time could facilitate laser self-starting. Ultrafast recovery times of few hundreds of femtoseconds have been reported making CNTs a suitable material for ultra-short pulse generation.

However, it is worth noting that, the recovery time is not so fast if we consider only isolated semiconducting CNTs, in this case, the recovery time would be in the order of 30 ps (Reich et al. 2004).

The ultrafast response time of CNTs depends on bundle and entanglement of both semiconducting and metallic CNTs. Excited electrons in semiconducting CNTs tunnel and couple to metallic CNTs, resulting in ultrafast recovery time and hence, the combination of the fast and slow processes within the CNT distribution, allows CNT to be a very efficient saturable absorbers. In addition, the inability to fully control the diameter of the CNTs during their growth is advantageous for the implementation of fiber lasers. In fact, the wide distribution of CNT diameters is responsible for the very broadband operation of CNT based saturable absorbers (Kivistö et al. 2009).

2.4.3.2 Third Order Nonlinearity

The first theoretical studies on the third order nonlinearity of CNTs appeared in 1998 when an estimated nonlinear refractive index coefficient, n_2, of as high as 2×10^{-12} m^2/W was predicted for the nanostructure (Margulis and Sizikova 1998). The theoretical value was several orders of magnitude higher than that of silica glass for instance with n_2 equal to 3×10^{-20} m^2/W and chalcogenide glass, with n_2 equal to 2×10^{-18} m^2/W which were receiving much attention as a suitable material for the fabrication of highly nonlinear fibers. The optical nonlinearity of CNT comes from the one-dimensional motion of the delocalized π-electrons at a fixed lattice ion configuration. These estimations attracted a great deal of attention since third order nonlinearity is responsible for phenomena such as third harmonic generation, optical Kerr effect, self-focusing and phase conjugation. As a result, third order nonlinear materials can be considered for optical functions such as optical switching, routing and wavelength conversion. The challenging part however, is that for most third order nonlinear optical materials, the third order susceptibility is too low for practical applications and hence the extremely high n_2 of CNT opened a new potential application for CNT (Yamashita 2011). Furthermore, in most cases there is a trade-off between high optical nonlinearity and its recovery. For instance, optical fibers generally exhibit low nonlinearity and a fast response time while semiconductor based devices generally operate with a high nonlinear coefficient and low response time. Remarkably, CNT devices offer a very high nonlinear coefficient combined with a fast response time due to their nanostructures and the Van der Waals induced bundling.

2.4.3.3 Optical Absorption Spectroscopy

When the optical properties of carbon nanotubes are referred, the focus is generally on the o absorption, photoluminescence, and Raman scattering of the nanomaterial. Optical absorption spectroscopy (OAS) measures the absorption of

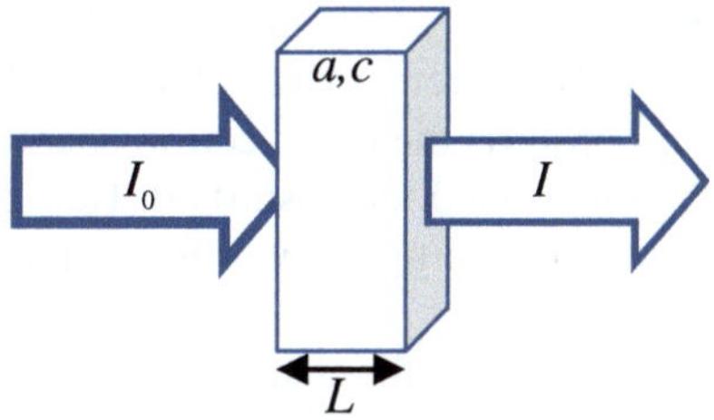

Fig. 2.6 Schematic illustration of Beer-Lambert absorption in a material

electromagnetic radiation which is a function of the frequency (or wavelength) of the incoming ray and its interaction with the sample. When light propagates through an absorbing medium, its intensity decreases exponentially according to the Beer-Lambert law, as

$$I(L) = I_0 e^{-aLc}, \tag{2.7}$$

where I_0 is the intensity of the incident light, a is the absorption coefficient, L is the optical path length, and c is the concentration of absorbing species in the material, as schematically illustrated in Fig. 2.6.

The absorbance for a given wavelength, A is then measured by obtaining the transmitted intensity relative to the incident intensity, given by the following expression,

$$A = -\ln\left(\frac{I}{I_o}\right). \tag{2.8}$$

The optical response of SWCNTs is dominated by transitions between valence and conduction bands. However, momentum conservation only allows transition between those pairs of singularities that are symmetrically placed with respect to the Fermi level. These vertical transitions are labeled as E_{ii} (i = 1,2,3,…), where basically, light will be absorbed when it is in resonance with the E_{ii} values for the (n,m) nanotubes in a sample. Despite the fact that a SWCNT has a sharply peaked density of states, the peaks that are found in OAS are broad due to sample heterogeneity (Tian et al. 2010). A collection of nanotubes however consists of many different types of (n,m) nanotubes mostly, each with a different set of absorption peaks, which add up to production of a broad peak that contains information about the diameter distribution in the sample. Thus, OAS of SWCNTs is often used to evaluate the mean diameter and diameter distribution of nanotubes in a sample, rather than their chiralities.

Although simple one-particle pictures are useful models to interpret some aspects of the experimental results (Liu et al. 2002), it has become increasingly clear that Coulomb interactions also play an important role in determining the optical transition energies of SWCNTs. Both theoretical calculations and experimental measurements show that the exciton binding energies are anomalously large at about 1 eV, in contrast to 10 meV energies that are commonly found in

3D material, which indicates the importance of many-body (MB) effects in the quasi-1D systems (Dresselhaus et al. 2005). The electron-electron Coulomb repulsion upshifts the one-electron energy and the electron-hole Coulomb attraction, on the other hand, downshifts it. Hence, the effects of many-body interactions on the transition energy are worth considering in modeling optical transition energies of SWCNTs.

Aside from the many-body corrections, the curvature effect and the C–C bond length optimization are also missing from the conventional tight-bond (TB) models. It has been shown that the long-range interactions of the π orbitals are not negligible and the curvature of SWCNT walls yields a sp^2–sp^3 rehybridization (Liu et al. 2002). The curvature effect was included in the TB model by extending the basis set of atomic to orbitals to the orbitals that form the σ and π molecular orbitals, according to the Slater-Koster formalism (Saito et al. 1998). This extended tight-binding (ETB) model, calculated within the frame work of a density functional theory (DFT), which incorporates the TB transfer and overlap integrals as functions of the C–C interatomic distance and is considered as one of the most accurate methods for modeling tight-bond energies (Tian 2012). The method includes long-range interactions and the effects of bond-length variations within the SWCNT sidewall (Samsonidze et al. 2004).

2.4.4 Thermal Properties

Prior to discovery of the properties of carbon nanotubes, diamond was known as the best thermal conductor. The heat flow and thermal conductivity along main axes in carbon nanotubes however is obtained to be of about 3500 W m^{-1} K^{-1} (Pop et al. 2006), whereas values between 900 and 2300 W m^{-1} K^{-1} is reported for the thermal conductivity of diamond. The thermal conductivity is specifically large along the main axis of CNTs, due to the easy propagation of atomic vibrations down the tube. In the direction transverse to its axis, however, the nanotube is much less rigid and the thermal conductivity in that direction is about a factor of 100 smaller (Dresselhaus et al. 2004; Sinha et al. 2005). A SWCNT has a room-temperature thermal conductivity across its axis (in the radial direction) of about 1.52 W m^{-1} K^{-1}. As another major different between carbon nanotubes and other carbon materials such as graphite, thermal expansion of conventional CNTs with rather large diameters is known to be essentially isotropic, whereas it is highly anisotropic along the long axis in case of graphite (Ruoff and Lorents 1995). There is however, some disagreement into the exact nature of the thermal conductivity of carbon nanotubes and it is believed that the thermal conductivity of CNTs depends on the temperature and the large phonon mean free paths as well (Berber et al. 2000).

2.5 Manufacturing Techniques

CNTs have probably been around much longer before they were practically discovered, as they may have been produced during various carbon combustion and vapor deposition processes. The electron microscopy however was not advanced enough at that time to distinguish them. Nowadays, there are several methods for producing carbon nanotubes, some of which are more specifically utilized for commercial synthesizing of carbon nanotubes.

2.5.1 Arc Discharge

Nanotubes were observed in 1991 in the carbon soot of graphite electrodes during an arc discharge that was intended to produce fullerenes. The carbon arc discharge method is actually the most common and perhaps the simplest way to produce CNTs (Wilson et al. 2002). However, the unfavorable point is that it produces a complex mixture of components, and requires further purification to separate the CNTs from the soot and the residual catalytic metals present in the crude product.

As illustrated schematically in Fig. 2.7, this method creates CNTs through arc-vaporization of two vertical carbon rods or electrodes placed end to end, separated by approximately 1 mm, in an chamber that is usually filled with inert gas at low pressure (Wilson et al. 2002). The carbon nanotubes, ranging from 4 to 30 nm in diameter and up to 1 mm in length, are grown on the negative end of the carbon electrode that contains a small piece of iron as a catalyst in a direct current (DC) arc-discharge evaporation of carbon at 20 V (Rafique and Iqbal 2011).

Iijima used this method first in 1991 to produce carbon nanotubes (Iijima, 1991), however the method was employed with some variations to produce different ranges of CNT diameters, types and amounts by different scholars (Ebbesen and Ajayan 1992;

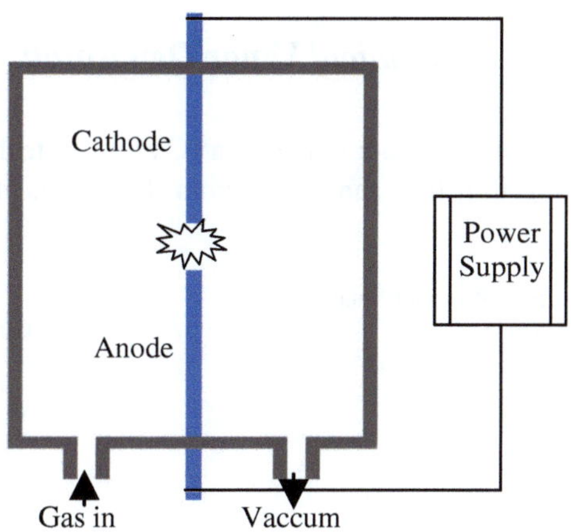

Fig. 2.7 Arch discharge chamber

Bethune et al. 1993; Journet et al. 1997). Investigations that are more recent have shown that it is also possible to create CNTs with the arc method in liquid nitrogen (Jahanshahi and Kiadehi 2013; Hosseini et al. 2012; Wang et al. 2005). As nanotubes were initially discovered using this technique, it has been the most widely used method of both single- and multi-walled nanotube synthesis.

2.5.2 Laser Ablation

Laser ablation process was first developed by Smalley research team in 1995 in Rice University (Guo et al. 1995). In the laser ablation process, as shown in Fig. 2.8, a graphite target is vaporized by being stroke by a pulsed laser in a high temperature reactor while an inert gas is bled into the chamber. The nanotubes are then formed on cool surfaces of the water-cooled Cu collector, as the vaporized carbon condenses (Rafique and Iqbal 2011).

The material produced by this method appears as a mat of "ropes", 10–20 nm in diameter and up to 100 μm or more in length. Each rope is found to consist primarily of a bundle of single walled nanotubes, aligned along a common axis. By varying the growth temperature, the catalyst composition, and other process parameters, the average nanotube diameter and size distribution can be varied (Rafique and Iqbal 2011). Arc-discharge and laser vaporization are currently known as the major approaches for obtaining small quantities of high quality CNTs. However, both methods are high-energy-consumer and involve with evaporating a limited amount of a carbon source, like graphite. Therefore, scaling up the production I these two techniques to the industrial level has not been achieved. In addition, by vaporization methods, CNTs grow in highly tangled forms, mixed with unwanted forms of carbon and/or metal species. Hence, purifying the produced CNTs, manipulating, and assembling them for building nanotube-device architectures for practical applications would be another major concern in these techniques.

2.5.3 Chemical Vapor Deposition

Chemical vapor deposition (CVD) is a technique that was majorly used in producing thin film semiconductors. However, in 1996 the technique was employed for

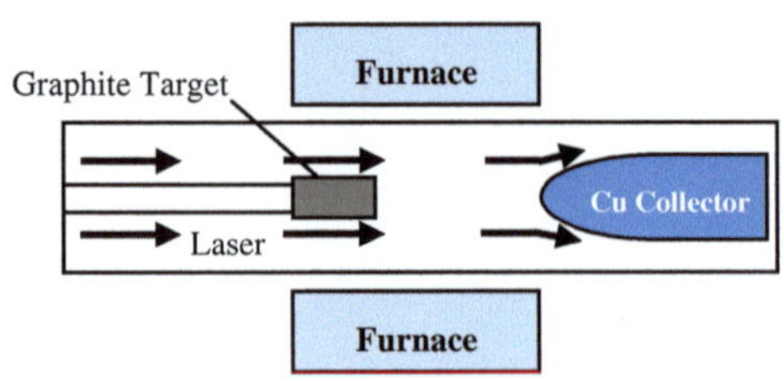

Fig. 2.8 Laser ablation reactor

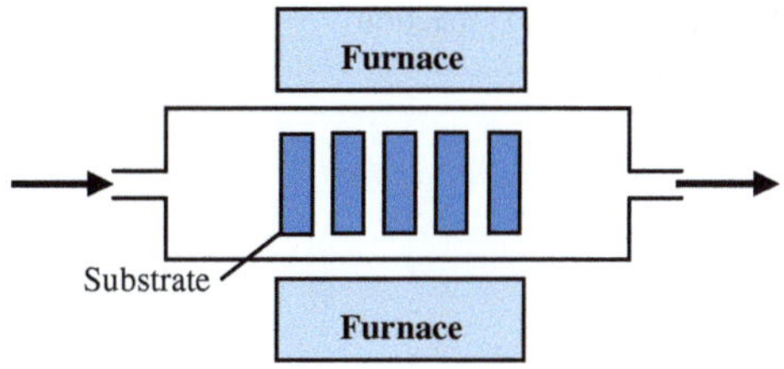

Fig. 2.9 CVD reaction chamber

large-scale production of carbon nanotubes (Li et al. 1996). The priority of this chemical CNT production technique is that the growth direction in large-scale production can be controlled over a substrate. In this method, a mixture of hydrocarbon or source gas and a process gas is made to react in a reaction chamber over a heated metal substrate. The hydrocarbon gas consists of ethylene, methane or acetylene and the process gas includes ammonia, nitrogen, and hydrogen). The temperature of the substrate is normally around 700–1000 °C, at atmospheric pressures. As a result of decomposition of the hydrocarbon gas, the CNTs are formed, deposit and grow on metal catalyst on the substrate. Figure 2.9 shows a thermal CVD

The catalysts particle and the substrate modification can result in production of different types of tubes. For instance, the nanotube diameter depends on the catalyst particle size. The substrate is usually silicon and hence, porous silicon is an ideal substrate for growing self-oriented nanotubes on large surfaces. In addition, glass and alumina are other materials used as the substrate material. The catalysts, on the other hand can be metal nanoparticles, like iron, cobalt, nickel or an alloy of the three catalytic metals. The catalysts are normally deposited on substrates by means of electron beam evaporation, physical sputtering or solution deposition. As the size of the catalyst affects the tube's diameter directly, the catalyst deposition technique should be opted carefully to yield desired results (Rafique and Iqbal 2011).

This method is also known as thermal CVD or catalytic CVD and has been used for decades for producing carbon nanotubes as well as other various carbon materials such as carbon fibers and filaments. However, some other techniques for the carbon nanotubes synthesis with CVD have been developed, such as plasma enhanced CVD, alcohol catalytic CVD, vapor phase growth, aero gel-supported CVD and laser-assisted CVD, some of which will be explained below.

2.5.4 Plasma Enhanced Chemical Vapor Deposition (PECVD)

The plasma enhanced chemical vapor deposition (PECVD) method, is a surface mediated growth technique, intended to generate vertically aligned tubes for commercial manufacturing. In this method, a glow discharge is generated in the reaction chamber by a high frequency voltage applied to its two parallel electrodes.

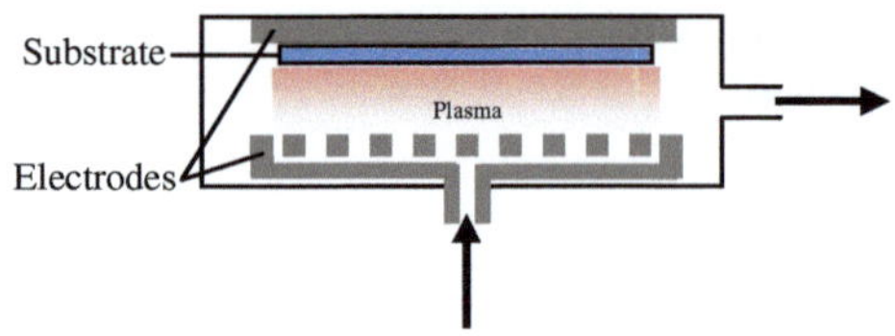

Fig. 2.10 PECVD reaction chamber

Figure 2.10 shows a schematic diagram of a typical plasma enhanced CVD apparatus with a parallel plate electrode structure.

In this method, substrate is placed on the grounded electrode and the source gas is supplied from the opposite plate. Catalytic metals, such as Fe, Ni and Co are used on for example the substrate to grow carbon nanotubes on them by the glow discharge generated from high frequency power. A carbon source gas, such as C_2H_2, CH_4, C_2H_4, C_2H_6, CO is supplied to the chamber during the discharge. Similar to Thermal CVD, the catalyst plays an important role in determining the nanotubes' diameter, growth rate, wall thickness, morphology and microstructure (Wilson et al. 2002).

2.5.5 *Vapor Phase Growth*

Vapor phase growth is another variant of the CVD technique in which the carbon nanotubes are produced directly from supplying reaction gas on the catalytic metal in the chamber without a substrate (Ge and Sattler 1994). In this method, two furnaces are placed in the reaction chamber and ferrocene is used as the catalyst. The first furnace is responsible of vaporization of catalytic carbon at a relatively low temperature and fine catalytic particles are formed. When they reach the second furnace, decomposed carbons are absorbed and diffused to the catalytic metal particles. The diameter of the carbon nanotubes produced by vapor phase growth are in the range of 2 to 4 nm for SWCNTs and between 70 and 100 nm for MWCNTs and the method has been employed successfully for large scale production of carbon nanotubes (Wei 2002; Deck and Vecchio 2005). Furthermore, recent investigations show that by means of this technique, the chirality of the produced CNTs is controllable (Liu et al. 2012).

2.5.6 *High Pressure Carbon Monoxide Reaction Method*

HIPCO® CVD, i.e. high pressure carbon monoxide chemical vapor deposition technique was introduced by Dr. Smalley's research team in 1998 (Nikolaev et al. 1999). The method involves introducing the catalyst in gas phase to the reaction chamber and rapidly mixing a gaseous catalyst precursor (such as iron carbonyl) with a flow of carbon monoxide gas at high pressure and high temperature. The catalyst precursor decomposes and nanometer-sized metal particles are formed. These tiny metal particles serve as a catalyst. On the catalyst surface, carbon monoxide molecules react to form carbon dioxide and carbon atoms, which bond together to form carbon nanotubes. This process selectively produces 100 % single walled carbon nanotubes. These carbon nanotubes can be used in electronics, biomedical applications and fuel cell electrodes (Unidym-Technology 2008).

As both the catalyst and the hydrocarbon gas are in the gas phase and the nanotubes are free from catalytic supports, the reaction can be operated continuously and hence, the method is suitable for large-scale synthesis of CNTs. Another variant of the technique was also introduced in 2001 in which a mixture of benzene and ferrocene, $Fe(C_5H_5)_2$ reacts in a hydrogen gas flow to form superconductive SWCNTs (Tang et al. 2001). In 2002, the method known as CoMoCAT® process was also introduced in University of Oklahoma to modify the process by using Cobalt and Molybdenum catalysts and CO gases. In this method, SWCNT are grown by CO disproportion at 700–950 °C, at a total pressure that typically ranges from 1 to 10 atm. This process was capable of growing a significant amount of SWCNTs (about 0.25 g-SWCNT/g-catalyst) in a couple of hours (Resasco et al. 2002). However, The CNTs produced by the HiPco process indicate a greater variety of diameters than the material produced by CoMoCAT Process.

2.5.7 Conclusion

In general, chemical CNT fabrication techniques have many advantages compared to arc discharge and laser ablation. In these methods, reaction processes are generally simple and easy to control and manipulate. In addition, raw materials are abundant and in form of gases, therefore, huge amount of energy would not be needed to process. As the methods are capable of producing vertically aligned CNTs on substrate, post refining and purification processes are minimized largely. Finally, large scale industrial manufacturing of CNTs is possible using chemical operations.

Other than the mentioned major methods, some other techniques have also been employed for manufacturing CNTs, namely helium arc discharge method (NASA 2005), electrolysis (Kaptay and Sytchev 2005), and flame synthesis (Wal et al. 2001) that have their own advantages and disadvantages and are only mentioned here to make them available for further studies.

2.6 Defects of Carbon Nanotubes

Carbon nanotubes have shown to be promising candidates as outstanding engineering materials, for their remarkable mechanical, thermal, optical and electrical properties. Therefore, they are expected to contribute in many outstanding applications in the world of science and industry. Nonetheless, there are several types of imperfections in the structure of carbon nanotubes, which may occur during the process of manufacturing or purposely introduced to the structure for specific applications, making functionalized carbon nanotubes. Defects are an inevitable part of all devices. Although in the macro world, the existence of minor defect does not influence the functionality of devices extensively, in the world of nano, even minor defects can alter the basic properties of the device significantly. Hence, the study of the defects is a major topic that should be investigated very carefully in nanoscience and technology. Therefore, point defects such as vacancies (Rahmandoust and Öchsner, 2009; Ghavamian et al. 2012; Ziaee 2013; Fakhrabadi et al. 2014), improper location of carbon atoms, e.g. perturbated or deformed structures (Rahmandoust and Öchsner, 2009; Ghavamian et al. 2012; Pullen et al. 2005), rearrangement of atoms in a new form, such as the Stone–Wales defect (Stone and Wales 1986; Nardelli et al. 1998); curving and bending of the CNTs (Yengejeh et al. 2014; Ray and Kundalwal 2013; Farsadi et al. 2013) and the presence of doped foreign atoms (Rahmandoust and Öchsner 2009; Ghavamian et al. 2012; Kaw 2006; Hernández et al. 1999b) are some of the defects involved with the structure of carbon nanotubes.

These defects may be majorly categorized into two main groups, i.e. topological defects occurring in the structure of carbon nanotubes, compared to the ideal arrangement of atoms, as well as defects that arise due to the doping of foreign atoms to the structure of carbon nanotubes. Introduction of any defects to the ideal structure of carbon nanotubes will result in a change in physical properties. These changes are in many cases desirable, namely for producing specific enhanced conductivity or semiconductivity or for boosting the sensing ability of the tube, where a controlled introduction of defects is required. On the other hand, the consequences of the existing defects may have undesirable influence on the properties of carbon nanotubes as well, namely decrease of mechanical strength of the structure, or lowered conductivity, that should be considered before the nanostructure is being applied as a functioning part of a device.

2.6.1 Topological Definition and Defects

The first main category of defects happens in the structural arrangement of carbon atoms. In this case, no foreign atoms are introduced to the structure and the defect rises from any imperfection occurring in the perfect hexagonal array of carbon atoms of the nanotube.

2.6.1.1 Atom Vacancies

A vacancy defect is a point defect that occurs naturally in the ideal hexagonal lattice of CNTs, as shown in Fig. 2.11a, in which one or several atoms are missing

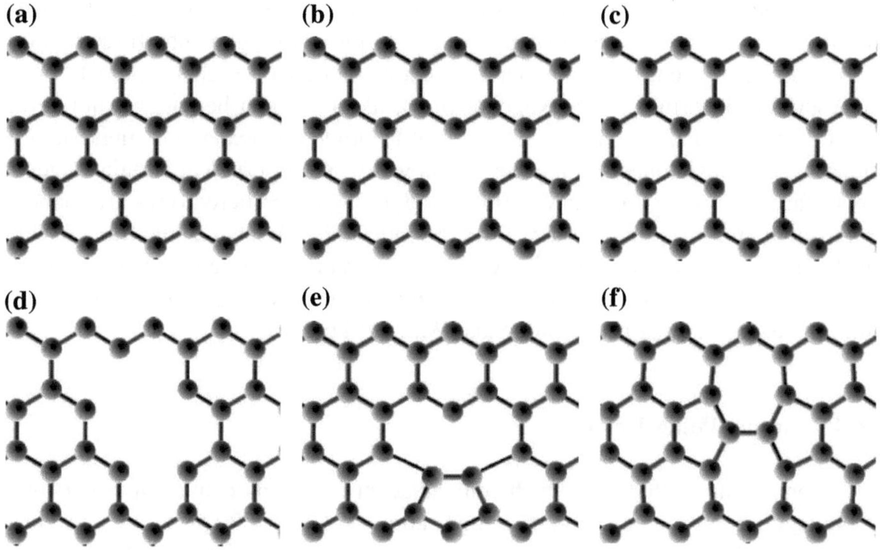

Fig. 2.11 **a** Ideal structure, **b** monovacancy (3DB), **c** double-vacancy, **d** triple-vacancy, **e** reconstructed monovacancy (5-1DB), and **f** Stone–Wales defect

from the structure. Vacancies, as illustrated in Fig. 2.11b through 2.11d, may be formed during solidification process, as a result of the vibrations of atoms, or they may happen due to local rearrangement of atoms after plastic deformation, or by heavy ion bombarding and irradiation of high-energy beams (Collins 2009; Ehrhart 1991). Once that the atom is knocked out its lattice site, the dangling bonds in the vacant site rehybridize to create to some new bonds (Ehrhart 1991) and the atom returns back to the surface of the structure to participate in a new bonding arrangement. Initial reconstruction of the three dangling bonds (sometimes referred to as 3DB defect) is in a way that two of the three free bonds, in a mono vacancy defect, bridge to form a strained 5-membered pentagonal ring, leaving a single dangling bond, hence abbreviated as 5-1DB. The 5-1DB defect is shown in Fig. 2.11e.

The theory of rehybridization which was initially proposed in 1998 by Ajayan et al. (1998) and was later confirmed and by Lu and Pan (2004) will be explained a bit more in detail below. However, the vacancies in the CNT even after initial rehybridization are energetically unfavorable. Hence, in many cases, they split into pentagon-heptagon or 5-7-7-5 defects, as shown in Fig. 2.11f. This type of defect, also known as the Stone–Wales defect, is the most prevalent type of topological defect in the structure of CNTs (Collins 2009).

If single-walled carbon nanotubes are exposed to high-energy beam irradiation, they will continuously loose atoms and try to reconstruct their structure (Ajayan et al. 1998). Uniform irradiation creates vacancies, which could cluster into larger vacancies in the structure, which are as explained before, energetically unstable.

This instability arises from the existence of dangling bonds. However, the structure tries to resolve the vacant sites through atomic rearrangements. Investigations of Ajayan et al. demonstrated that the reconstruction of vacancy defects occur successfully by forming nonhexagonal rings, like pentagon-heptagon pairs defect in the nanotube lattice (Ajayan et al. 1998). Other nonhexagonal rings such as squares, octagons, nonagons, and decagons were also observed at certain stages of the surface reconstruction (Kaw 2006). Though, high-membered rings are found to be unstable and thus, they disappear by the mechanism known as the Stone–Wales mechanism. As a result of the Stone–Wales mechanism, the constituent members of the stable rings in the structure of carbon nanotubes will be confined to a maximum of seven-member rings (Stone and Wales 1986).

2.6.1.2 Stone–Wales Defect

The Stone–Wales defect, in which four adjacent hexagons convert into two pairs of pentagon-heptagons or 5-7-7-5 configuration (Pozrikidis 2009; Stone and Wales 1986) is the final stage in atomic reconstruction of vacant sites. The Stone–Wales defect, also known as the pentagon-heptagon defect, is the most abundantly detected defect in the structure of carbon nanotubes that occurs naturally after initial formation of vacancies. This type of defect is expected to influence the electrical (Azadi et al. 2010), optical (Zhou et al. 2014), thermal (Wei et al. 2012) and mechanical properties (Pozrikidis 2009; Xiao et al. 2010) of carbon nanotubes, to different extents. However, FEM simulation results showed that although the introduction of defects to the body of CNTs weakens their elastic strength, the system is almost capable of restoring its original high elastic modulus by reconstructing its atoms in form of the 5-7-7-5 or Stone–Wales arrangement (Rahmandoust and Öchsner 2015). The Stone–Wales defect can be imagined to be formed by an in-plane $\pi/2$ rotation of a C–C bond (Pozrikidis 2009), as schematically illustrated in Fig. 2.12.

Other than the vacancies that happen naturally and the associated defects that arise from reconstruction of the vacant sites, there are some other defects that perturbates the ideal hexagonal arrangement of atoms in carbon nanotubes. These defects, as explained below, are mostly happening due to intentionally introduced changes in the structure seeking a specific application of CNT as a part of a device.

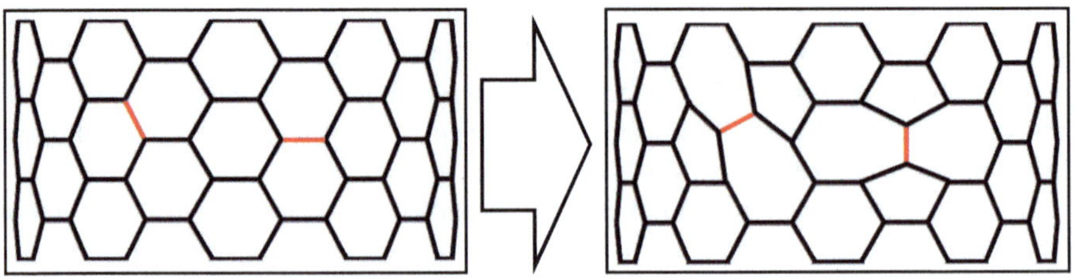

Fig. 2.12 Schematic illustration of the bond rotation in Stone–Wales defect

2.6.1.3 Kink Junctions

In order to produce molecular-size metal-semiconductor, metal-metal, or semiconductor-semiconductor nanojunctions in carbon-based nanoelectronics, some intramolecular connections are formed between two CNTs that have different chiralities (Saito et al. 1996; Chico et al. 1996). The tubes in turn should be connected to each other in a way that the strain energy is minimized and hence one or multiple topologic pentagon or heptagon defects are interposed into the structure to cover the chirality gap, creating a kink junction or a knee connection (Scarselli et al. 2012; Yao et al. 1999), as illustrated in Fig. 2.13. These topological changes create only a small local deformation in the width of the nanotubes (Kaw 2006).

Other than the targeted electrical influence of this type of heterojunction defect, the topological change affects the mechanical properties of the tubes as well. The kink intramolecular junctions tend to influence the stiffness, shear strength and buckling behavior of the carbon nanotubes (Yengejeh et al. 2013; Yengejeh et al. 2014).

2.6.1.4 Welding CNTs

Welding of carbon nanotubes by means of electron beams at elevated temperatures is another technique of merging nanostructures covalently. These molecular junctions are capable of producing various stable geometries similar to X, Y, and T, as shown in Fig. 2.14, that can act as multi terminal electronic devices (Terrones et al. 2002; Ponomareva et al. 2003).

The studies on the density of state in these hetrojunctions show that at the region of welding the one-dimensional characteristics of the tube are dramatically affected, which is an important factor in electronic device applications and controlled fabrication of nanotube-based molecular junctions and network architectures that are expected to exhibit exciting electronic and mechanical properties (Kaw 2006).

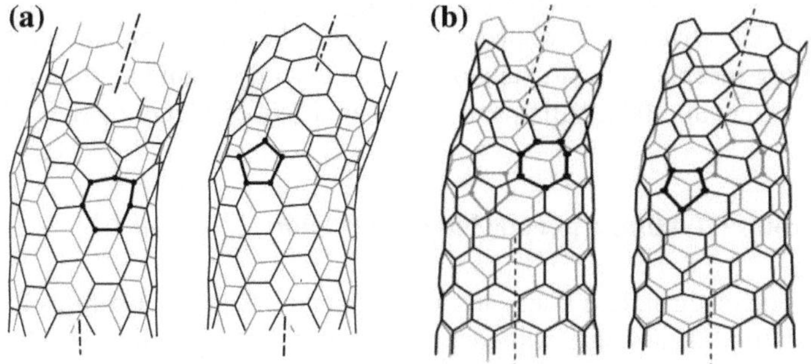

Fig. 2.13 Knee connection between two different CNTs with **a** similar, and **b** different diameters

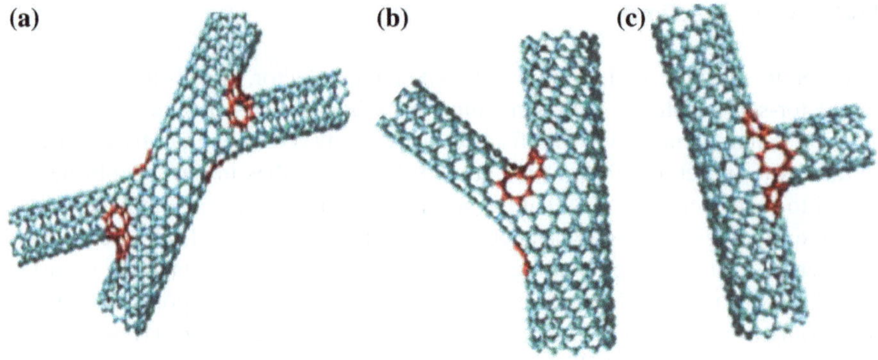

Fig. 2.14 Welded carbon nanotubes with **a** X-, **b** Y-, and **c** T-geometries (adapted from Terrones et al. (2002), with permission; Copyright 2002 by The American Physical Society)

2.6.2 Substitutional Doped Defect

Doping of foreign non-carbon atoms into the structure of sp^2 carbon nanotube systems, provides the opportunity of tailoring the electronic, vibrational, chemical and mechanical properties (Jorio et al. 2008; Terrones et al. 2008). It is in fact an efficient way to modify these nanostructures for different emerging applications. The substitutional heteroatom, like boron, silicon and nitrogen, can reach relatively high doping levels of about 2 % or higher (Wei et al. 2009), which results in altered properties. For instance, regardless of the chirality of the carbon nanotube, both boron- and nitrogen-doped carbon nanotubes show metallic behavior, although the states of acceptors (p-doping) or donors (n-doping) are strongly localized near the Fermi level (valence band) in both cases (Carroll et al. 1998; Czerw et al. 2001). On the other hand, the existence of high percentages of silicon, nitrogen and boron atoms lowers the stiffness of carbon nanotubes remarkably, although the Young's modulus remains above 0.5 TPa (Rahmandoust and Öchsner 2009; Ghavamian et al. 2012). The electron or hole-rich structures that can be obtained by doping group V or III atoms, respectively, can be used to produce n- or p-type nanotubes, which make valuable molecular heterojunction devices (Choi et al. 2000). The structure of doped carbon nanotubes had been studied to observe the pertaining morphological changes compared with the undoped counterparts of each tube (Gai et al. 2004; Villalpando-Páez et al. 2006; McGuire et al. 2005).

2.6.2.1 Group III Atoms; p-Type Nanoconductor

Boron has one electron less than carbon and hence, when it substitutes for C atoms inside the lattice of a single-walled carbon nanotube, a localized state below the Fermi level appears. The absence of an electron in other words, creates a hole in the structure, and the tube converts into a p-type nanoconductor. From the chemical absorption point of view, this structure would be therefore more likely to react

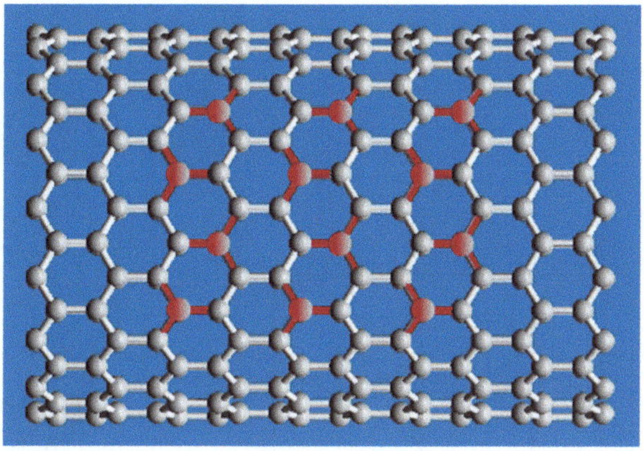

Fig. 2.15 Schematic illustration of boron-carbon islands. Boron atoms and the B–C bonds are marked with *red*

with donor-type molecules, which have electrons to share. Figure 2.15 shows boron-carbon islands in a single walled carbon nanotube.

Boron doped CNTs were produced by growing carbon nanotubes by arc discharge, using a Boron-rich consumable anode (Carroll et al. 1998). Later, the McGurie team produced B-doped single-walled carbon nanotubes with controlled low concentration of dopants (McGuire et al. 2005). Arc Discharge, on the other hand, is mainly used to fabricate B-doped multi-walled carbon nanotubes (Blase et al. 1999). Doping of boron atom during nanotube growth influences the chirality of the produced B-doped CNT. In other words, the introduction of a boron source in the synthesis process will control the tube chirality, toward a preferred zigzag or near zigzag chirality with $\pm 3°$ twist angle (Blase et al. 1999; Hsu et al. 2000).

2.6.2.2 Group IV Atoms; Hillocks

Silicon could also be doped inside SWCNTs (Baierle et al. 2001). Theoretical studies show that substitutional Si atoms induce a strong outward deformation, creating hillocks, making the surface more reactive compared to undoped CNTs (Baierle et al. 2001; Fagan et al. 2003; Rahmandoust and Öchsner 2009). Other than making the nanostructure readier to participate in surface reaction with the surrounding environment, based on finite element studies, the implementation of doped silicon inside the lattice of CNTs results in a linear reduction of the structure's elastic modulus, both in single- and multi-walled carbon nanotubes (Rahmandoust and Öchsner 2009; Ghavamian et al. 2012). Figure 2.16 shows the hillocks created as a result of Si substitution.

2.6.2.3 Group V Atoms; n-Type Nanoconductor

Substitution of group V atoms, like phosphor and nitrogen, with one extra valance electron compared to carbon, induces localized states that may be both below and above the Fermi level. It means that for instance, in case of N doping

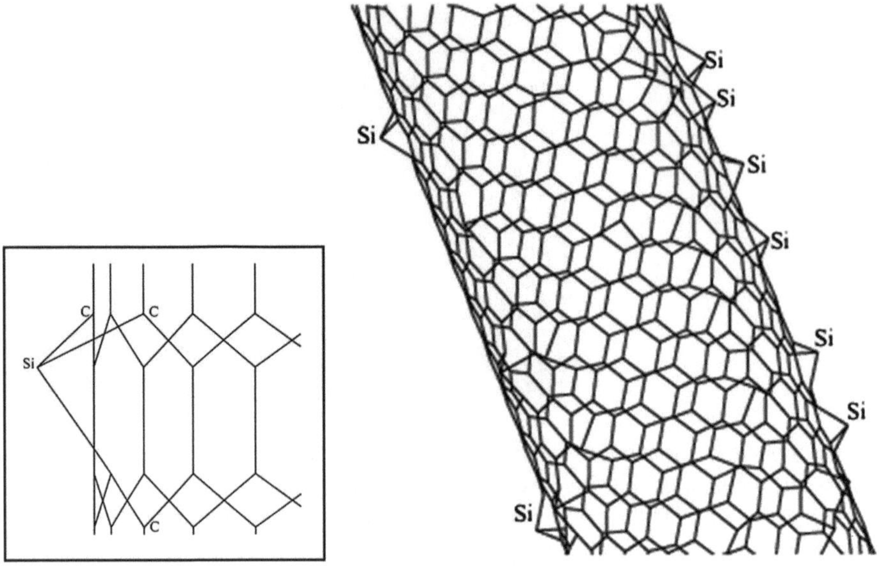

Fig. 2.16 Si-doped SWCNT

in CNTs, also known as CNx nanotubes, both n- and p-type conducting behavior are expected, depending on the type and the level of doping inside the hexagonal sheet. As shown in Fig. 2.17, there are three different configurations for nitrogen when replacing carbon in an sp^2 lattice (Terrones et al. 2012), namely substitutional graphitic nitrogen, pyridine-like nitrogen, i.e. type1 and type2, respectively, and pyrrole-like nitrogen, type3, from which only the first two are stable in carbon nanotubes, which produce n-type and p-type conductors respectively (Terrones et al. 2008). Furthermore, it has been observed that the number of graphitic nitrogens compared to pyridine-like Ns increases as the overall content of doped nitrogen increases (Terrones et al. 2008).

Nitrogen-doped carbon nanotubes can be fabricated by thermal decomposition of N-containing hydrocarbons over metal particles. The first report on the formation of aligned arrays of N-doped MWCNTs, with concentrations below 2 %, was in 1999 (Terrones et al. 1999). Later in 2001, a similar technique was applied to produce Nitrogen-doped SWCNTs at elevated temperatures (Czerw et al. 2001). More advanced methods were proposed in 2005 and 2006 for fabricating isolated N-doped CNTs and long stranded bundles of defective structures, respectively (Keskar et al. 2005; Villalpando-Páez et al. 2006). However, fabrication of highly ordered structures containing large concentrations of N is hard to achieve (Sen et al. 1997) but on the other hand, low concentrations of doped materials are difficult to detect and hence, the detection of the amount of dopants was only possible using Raman spectroscopy (McGuire et al. 2005).

In case of continuous growth of N-doped single walled carbon nanotubes, only substitutional graphitic nitrogens are observed, in which N is connected to three

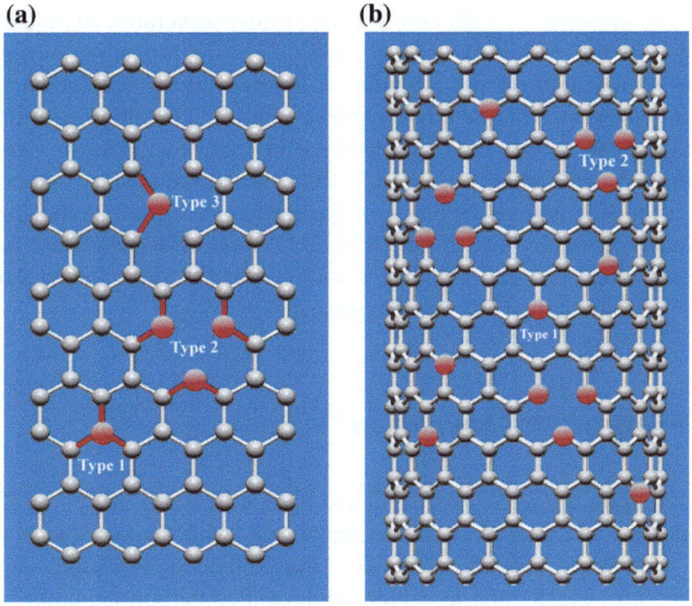

Fig. 2.17 Possible configurations of nitrogen-carbon bonding in **a** Graphene sheet and **b** Carbon nanotube

carbons and has n-type characteristics. Substitutional graphitic nitrogens are more likely to react with acceptor molecules (Villalpando-Páez et al. 2006; Terrones et al. 2008). Pyridine-like N structures, on the other hand, can occur in SWCNTs provided that an additional carbon atom is removed from the lattice. In this type of defect, each N atom is bonded to two carbon atoms, responsible for creating cavities and corrugation in the nanotube structure, which is responsible for the metallic behavior of N-dope CNTs (Czerw et al. 2001).

The mechanical behavior N-doped carbon nanotubes are also studied both experimentally and computationally. The Young's modulus for N-doped MWCNTs is as low as about 30 GPa, due to the relatively high N concentration. The high percentage of defect, i.e. of about up to 5 %, lowers the mechanical strength of CNTS significantly. However, if the low concentration of about 0.5 % nitrogen is doped, the elastic modulus of the structure will not change substantially (Gao et al. 2000).

2.7 Modeling of Carbon Nanotubes

The finite element method has been successfully used as a practicable, powerful, and cost-effective analysis method for the numerical study of a wide range of engineering problems. FEM has many applications from deformation and stress analysis of automotive, aircraft, building, and bridge structures, to field analysis of heat, fluid flow, magnetic flux, leakage, and other flow problems. In this method, a complex region defining a continuum is divided into simple geometric forms called *finite elements*. The material properties and the governing relationships are considered over these elements and declared in terms of unknown values at element corners. An assembly process, duly considering the loading and constraints, results in a set of equations. Solution of these equations gives us the approximate behavior of the continuum (e.g. Tirupathi and Ashok 1997). Software based tools have been therefore proven to be a very powerful and well-organized tool in characterizing and simulating CNT-based composites.

2.7.1 Nanostructural Simulation

The simulation on the nanoscale explained here, covers the modeling of different types of perfect and imperfect carbon nanotubes, with the purpose of achieving their key mechanical properties. Initially, the method of modeling the structure and properties of zigzag and armchair single-walled carbon nanotubes are introduced and then multi-walled carbon nanotubes are studied. As it is very common to encounter imperfect structures in experiments, modeling of the physical properties of imperfect structures will also be discussed.

The structure of single-walled carbon nanotubes is composed of several carbon–carbon bonds joined together in a hexagonal arrangement. In order to study the mechanical properties of SWCNTs, these bonds were successfully defined in FEM by Euler–Bernoulli linear thin elastic beams, shown in Fig. 2.18 in black color. This element represents the covalent bonding between two adjacent carbon atoms in the nanotube structure (Rahmandoust and Öchsner 2009).

When it comes to MWCNTs, as explained earlier, there are also non-covalent, van der Waals interactions between atoms from any of the two layers of carbon nanotubes. As this type of interaction does not have a material-based nature and hence, it is considered as a weak interaction, *spring* element, with a suitable definition of spring coefficient, was opted to replace it (Rahmandoust and Öchsner 2012a, b). A schematic view of the starting ring of a DWCNT is shown in Fig. 2.18, in which the radial red lines represent the above-explained spring elements and the surrounding black lines are the thin elastic beam elements, representing C–C bonds.

To obtain Young's modulus of the structure, for example, the produced CNTs were fixed from translation and rotation on one end, and a point load was applied to the other end. Furthermore, in order to characterize the properties of carbon nanotubes, using the finite element models, the mechanical properties of each carbon–carbon atomic bond are necessary to be obtained based on its chemical properties. These properties should be then defined as the material properties in the pertinent finite element model.

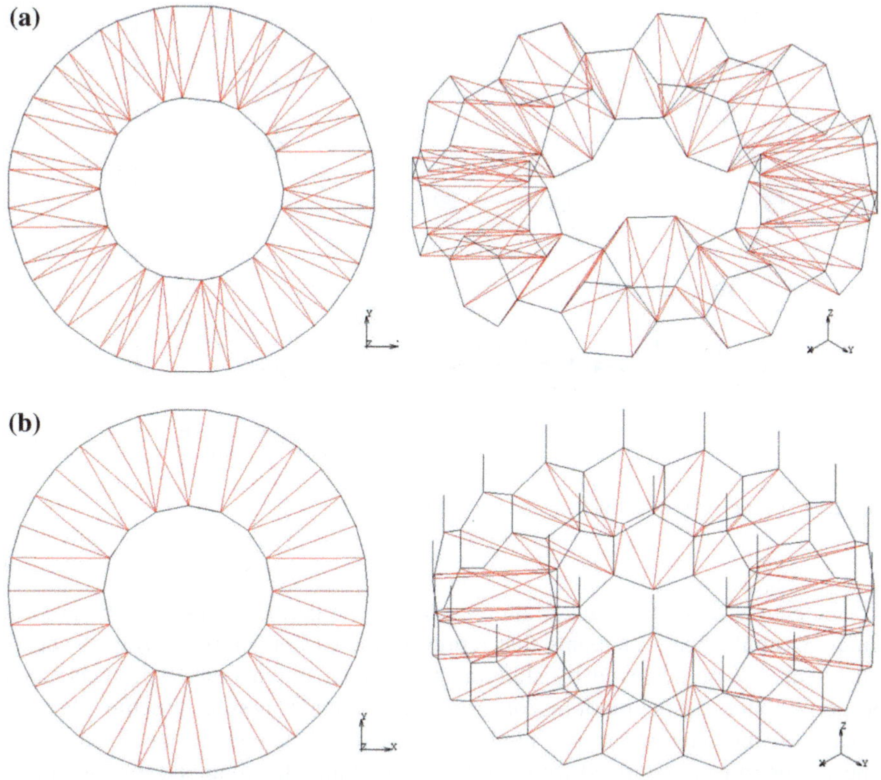

Fig. 2.18 *Top*- and *side*-view of first ring for **a** (10,10)-(5, 5) armchair DWCNT, and **b** (17, 0)-(8, 0) zigzag DWCNT

2.7.1.1 Properties of C–C Bonds

Each CNT is considered to be made up of unit blocks that are repeated and rotated around an imaginary circle with a diameter equal to that of the desired CNT. As explained before, the diameter of any specific CNT can be calculated by Eq. (2.2). This is only possible if the CNT unit block is bendable on its bond joints, i.e. on the location of carbon atoms, shown in Fig. 2.19 by dots, to let the structure cover the curvature angle of its tube diameter (Chang and Gao 2003). If we consider having straight non-bending carbon–carbon bonds as FE beam elements, then each unit block of an armchair and a zigzag CNT can be divided into four and two rigid parts, respectively, as shown in Fig. 2.19.

The structure can then be generated by producing the first ring by rotating and repeating the unit blocks around a circle with a diameter equal to that of the desired CNT. Later, after assigning the geometric and the material properties of bonds, these rings are copied and shifted up repeatedly, as shown in Fig. 2.20, to make the whole CNT structure, up to the desired height.

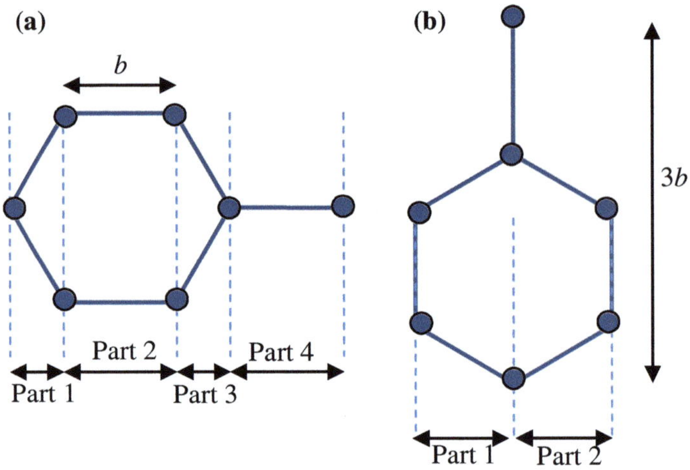

Fig. 2.19 Unit blocks of **a** armchair, and **b** zigzag SWCNTs

The geometric and the material properties of the three-dimensional thin, elastic Euler-Bernoulli beam elements that represent the carbon–carbon covalent chemical bonds can be achieved by studying the atomic interactions. The mechanical properties and elastic modulus of covalently bonded atomic structure depends greatly on the interatomic potential energy acting between the involved atoms. These properties are derived from the bond stretching, bending and torsion behavior. Young's modulus, shear modulus and Poisson's ratio are, for example, values that describe how an isotropic material behaves under different loads in the elastic range. If the Hooke's law is applied in the elastic range, for each generalized beam element, two of these values, i.e. Young's modulus, shear modulus and Poisson's ratio, are enough to be assigned. These quantities can be obtained for a carbon–carbon bond, from the equations given below (Li and Chou 2003; Kalamkarov et al. 2006).

The strain energy under pure axial load P (pure tension) is given by

$$U_P = \int_0^b \frac{P^2}{2EA} db = \frac{EA}{2b}(\Delta b)^2. \tag{2.9}$$

The strain energy of a beam element subjected to a pure bending moment M is

$$U_M = \int_0^b \frac{M^2}{2EI} db = \frac{EI}{2b}(\Delta \alpha)^2. \tag{2.10}$$

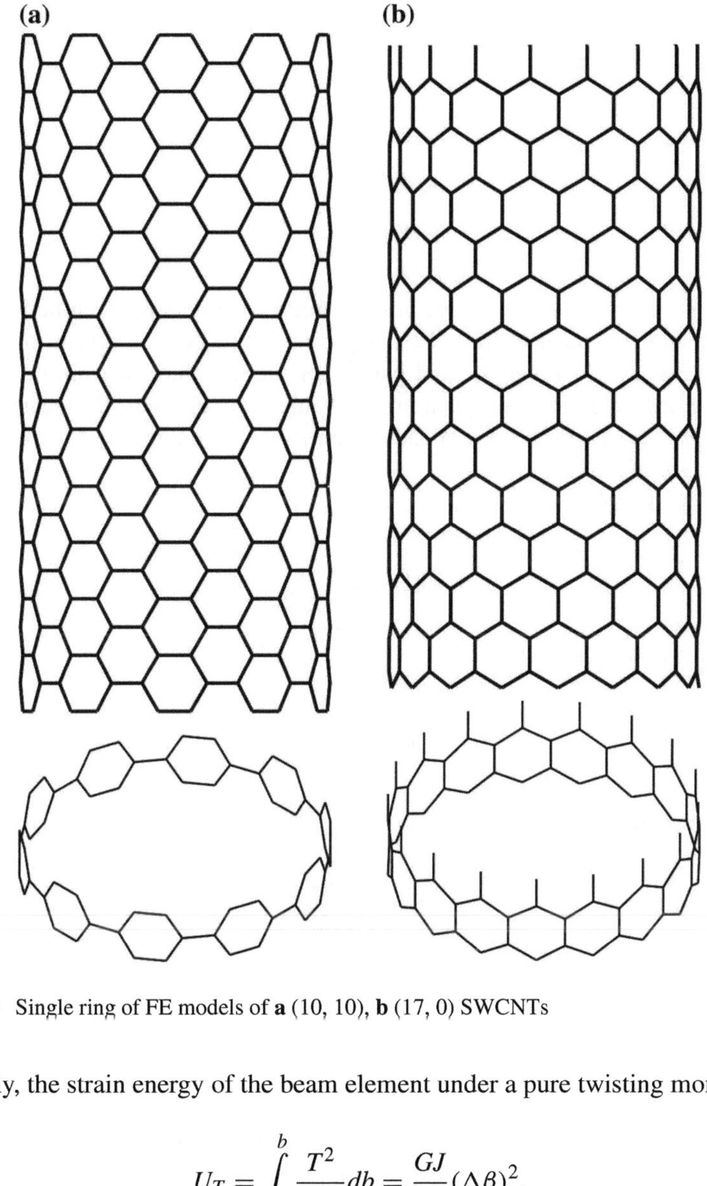

Fig. 2.20 Single ring of FE models of **a** (10, 10), **b** (17, 0) SWCNTs

Similarly, the strain energy of the beam element under a pure twisting moment T is

$$U_T = \int_0^b \frac{T^2}{2GJ} db = \frac{GJ}{2b} (\Delta \beta)^2. \tag{2.11}$$

In the above equations, assuming the terms Δb, $\Delta \alpha$ and $\Delta \beta$ as the axial deformation, bending angle and twist angle respectively, one obtains the tensile stiffness (EA), bending stiffness (EI) and torsion rigidity (GJ) of the structural model in terms of the molecular mechanics force constants k_r, k_θ and k_φ as follows:

$$k_r = \frac{EA}{b}, \tag{2.12}$$

$$k_\theta = \frac{EI}{b}, \tag{2.13}$$

$$k_\varphi = \frac{GJ}{b}. \tag{2.14}$$

In the context of *molecular mechanics*, a force field refers to the functional form and parameter sets used to describe the potential energy of a system of particles or atoms. Force field functions and parameter sets are derived from both experimental studies and high-level quantum mechanical calculations. By means of these constants, the elastic properties of each beam element (assumed to have a diameter d_b and length b) are obtained as:

The bond diameter

$$d_b = 4\sqrt{\frac{k_\theta}{k_r}}, \tag{2.15}$$

Young's modulus

$$E = \frac{k_r^2 b}{4\pi k_\theta}, \tag{2.16}$$

Shear modulus

$$G = \frac{k_r^2 b k_\varphi}{8\pi k_\theta^2}, \tag{2.17}$$

and Poisson's ratio

$$\nu = \frac{k_\theta}{k_\varphi} - 1. \tag{2.18}$$

Based on above equations, for each elastic beam element or in other words for each C–C bond, Young's modulus will be $E = 5.484 \times 10^{-6}$ N/nm^2 and Poisson's ratio $\nu = 2.15$. The Poisson's ratio obtained here for each carbon–carbon bond, based on molecular mechanics calculations, is equal to 2.15, which does not seem to be an acceptable engineering parameter. This is because molecular mechanics constants were used here to calculate the Poisson's ratio of each bond by Eq. (2.18), which is actually based on classical mechanics and may not be applied at molecular level. Therefore, after encountering the same problem, Natsuki et al. suggested another method for the analytical calculation of Poisson's ratio of a C–C bond.

Table 2.1 Material and geometric properties of a C–C bond

Chemical bond:	Carbon–carbon covalent bond	
Corresponding force field constants	$k_r = 651.97$ nN/nm	
	$k_\theta = 0.8758$ nN nm/rad^2	
	$k_\varphi = 0.2780$ nN nm/rad^2	
Equivalent finite element type:	3-D thin elastic beam	
	Geometric properties	
	Bond radius	$R_b = 0.0733$ nm
	Cross section area	$A = 0.0169$ nm^2
	Second moments of area	$I_{xx} = I_{yy} = \pi R_b{}^4/4$ $= 2.26820 \times 10^{-5}$ nm^4 $I_{zz} = J = \pi R_b{}^4/2$ $= 4.53456 \times 10^{-5}$ nm^4

They obtained the value by the following equation equal to 0.27, based on the following equation.

$$\nu = \frac{b^2 k_r - 6k_\theta}{b^2 k_r + 18k_\theta}. \tag{2.19}$$

Table 2.1 provides the material and geometric properties of the finite element corresponding to a covalent carbon–carbon chemical bond.

2.7.1.2 Properties of Non-covalent Interactions

For modeling the CNT structure as a finite element model, covalent bonds are represented by thin elastic beam elements, but when it is about modeling multi-walled carbon nanotubes, it is necessary to define non-covalent forces acting between each two layers of atoms, or in other words, between each two walls of CNTs, as well. This kind of force is defined by the Lennard-Jones force, Eq. (2.4), and it is a function of the distance between the two interacting particles. Although the Lennard-Jones force is a week force, there is an influencing distance interval considered for the Lennard-Jones force acting between any two layers of MWCNT and that is when the distance between the carbon atoms is between 0.33 to 0.38 nm (Kalamkarov et al. 2006). Thus, it should be considered that by applying an external load on the structure, the locations of atoms with respect to each other change and as a result, the Lennard-Jones force changes. Since there is no material between the interacting particles, it is very important to find the most suitable element to represent this interaction.

Therefore, *spring element* was opted as the best choice, in the sense that it is acting between *two* nodes, it is of non-material nature and it follows a force-displacement equation. The only challenging part of this choice is that spring force

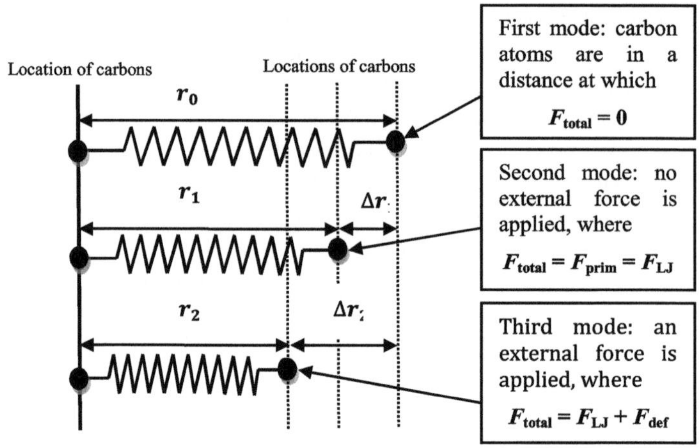

Fig. 2.21 Spring modeling of the Lennard-Jones force

is defined as a function of displacement and not distance, but the Lennard-Jones interaction is a function of distance between two interacting particles. Thus, based on the initial distance between each pair of interacting atoms, a force intersect like a pre-strained spring, has to be defined by a coordinate transformation. Any changes in the distance will later influence the interacting force according to the updated force-displacement equation. Figure 2.21 shows the described situation schematically.

As the Lennard-Jones force equation is a function of the inverse of distance, the software was incapable of processing it in order to derive the pertinent stiffness matrix. Considering that this complex force influences only a very small region of its whole diagram, in the case of MWCTs, two approximations of the force were developed initially to solve the problem. At the specific interval of 0.33–0.38 nm between two layers of coaxial tubes, the Lennard-Jones equation was approximated by a 5th and a 7th degree polynomial with norm of residuals equal to about 10^{-2} and 10^{-4} Joules, respectively. The diagram of the exact equation and its approximations are presented in Fig. 2.22. Although the 7th degree polynomial results in a less norm of residual, the 5th degree shows a better fitting in the critical area of around the initial distance between the two involved atoms, which makes it in general a more desirable choice.

The stiffness of the equivalent spring can also be approximated with a linear polynomial. For this purpose, two cases have to be considered. At first, there is no load applied on MWCNT and the structure is at rest. Thus, no deformation is imposed to the models. In this stage, the Lennard-Jones force between each two carbon atoms of two different walls of the MWCNT, at their initial distance, is referred to as F_{prim}. The second stage is when the model is exposed to an external force which causes the dislocation in the initial position of carbon atoms with

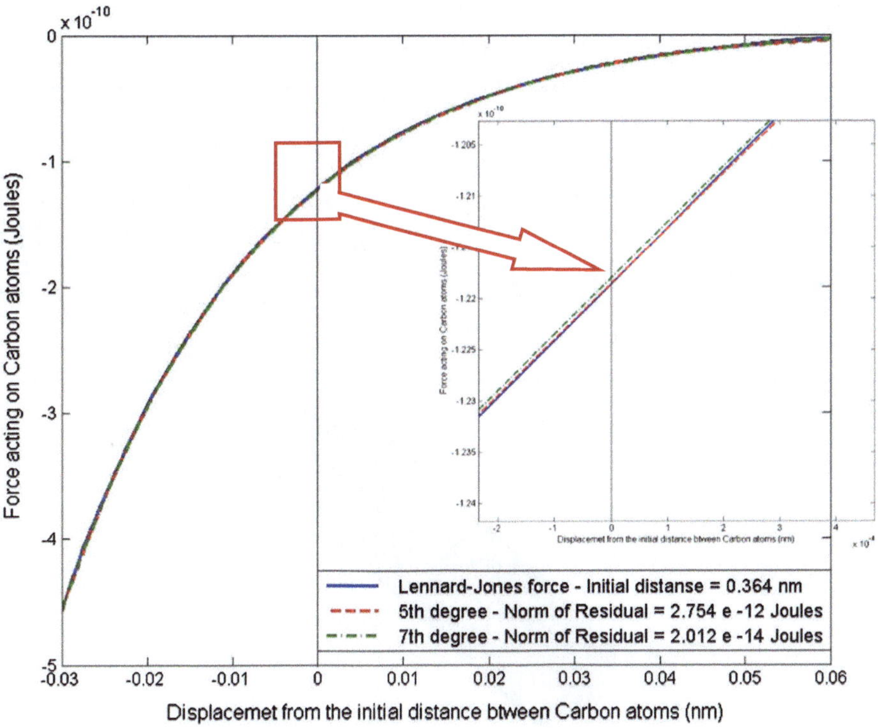

Fig. 2.22 Comparison of the Lennard-Jones force with its approximations

respect to each other and hence a change in the distance between the atoms, which consequently leads to a change in the Lennard-Jones force acting between the two interacting atoms. This force, is defined by $F_{\text{def}} = k \cdot \Delta r$ where k, the stiffness should be calculated approximately for this structure.

The obtained value for stiffness is approximately equal to 0.24245 N/m, calculated by choosing two points from the Lennard-Jones force diagram which is assumed to be linear in the assigned so called Lennard-Jones effective distance for MWCNTs, i.e. $0.33 < r < 0.38$ nm. Considering the discussion brought up before, Eq. (2.18) is obtained to comprehensively define the total force acting between each two carbon atoms, before or after deformation.

$$F_{\text{total}} = F_{\text{prim}} + F_{\text{def}} = \frac{4\varepsilon}{r}\left[-12\left(\frac{\sigma}{r}\right)^{12} + 6\left(\frac{\sigma}{r}\right)^{6}\right] + k \cdot \Delta r \qquad (2.20)$$

Figure 2.23 shows an armchair DWCNT unit block, an armchair 5-walled ring and a 5-walled CNT, in which the red lines, connecting the carbon atoms of two walls, represent the Lenard-Jones forces and the black lines are the C–C bonds.

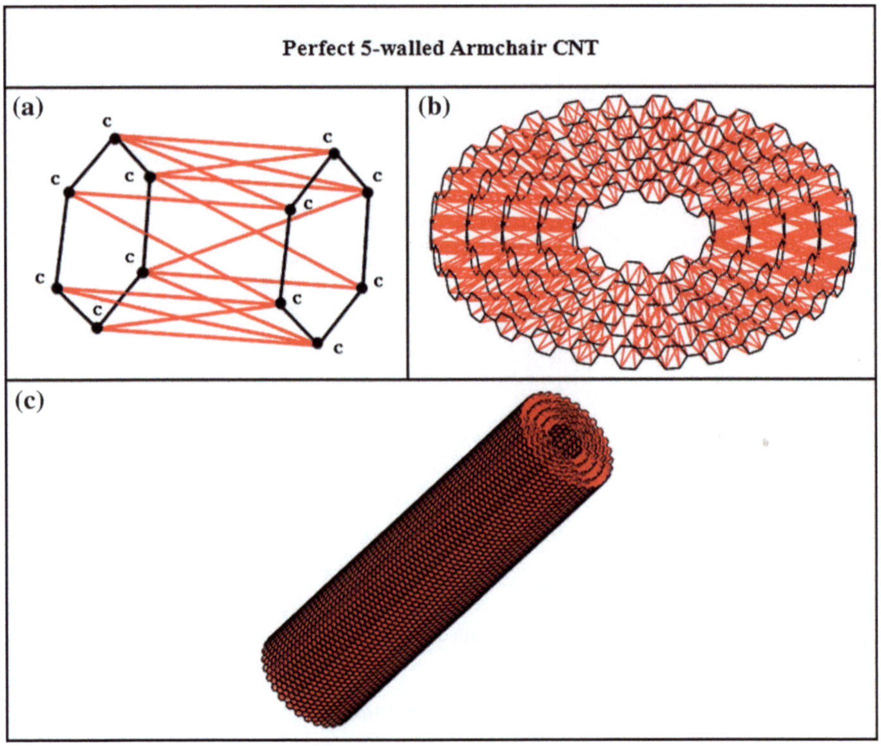

Fig. 2.23 **a** Armchair DWCNT unit block, **b** armchair 5-walled ring, and **c** armchair 5-walled (10, 10)-(15, 15)-(20, 20)-(25, 25)-(30, 30) CNT

2.7.2 Mechanical Properties of SWCNTs

Let us consider in the following a carbon nanotube, fixed on one end from translation and rotation, as it is schematically shown in Fig. 2.24, for two sample armchair and zigzag SWCNTs. In addition, a fixed displacement is applied on the other end along the tube's principal axis (here z-axis). As a result of this displacement boundary condition, a reaction force in z-direction is obtained at each node and the sum of these reaction forces equals the net reaction force P.

2.7.2.1 Young's Modulus

Using the net reaction force P and the deformations, one can find Poisson's ratio ν, Young's modulus E and Shear Modulus G of the structure using the Hooke's law in the elastic range as follows:

$$E = \frac{\sigma}{\varepsilon}, \tag{2.21}$$

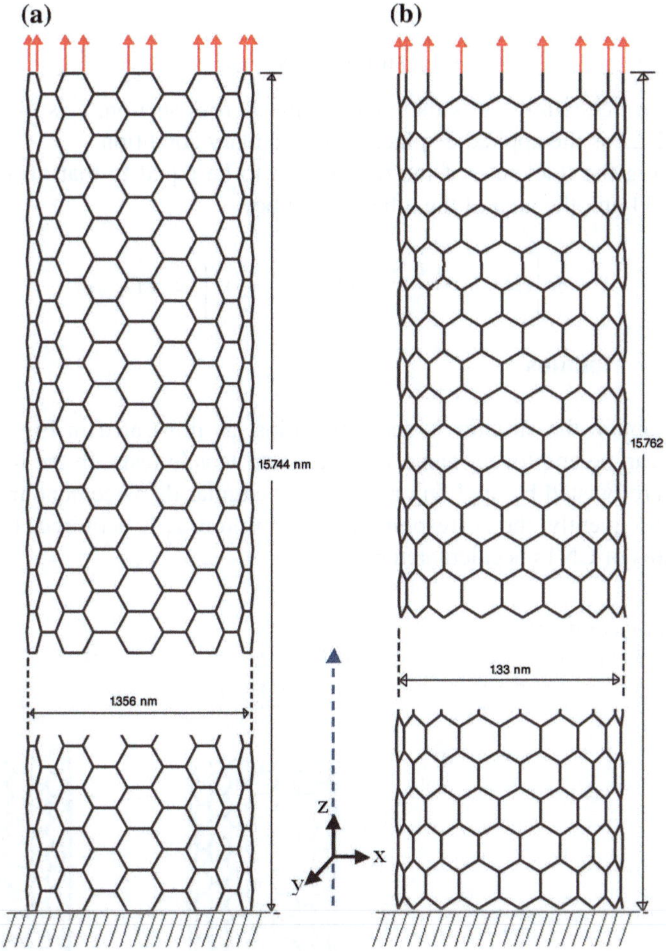

Fig. 2.24 Geometrical dimensions and applied boundary conditions on **a** (10, 10) armchair and **b** (17, 0) zigzag single-walled carbon nanotube

$$\nu = -\varepsilon_{\text{lateral}}/\varepsilon_{\text{axial}} = -\frac{\Delta D}{D}\Big/\frac{\Delta L}{L}, \tag{2.22}$$

$$G = \sigma_{\text{lateral}}/\sigma_{\text{axial}} = \frac{E}{2(\nu + 1)}, \tag{2.23}$$

where

$$\sigma = \text{ stress } = P/A, \tag{2.24}$$

and

$$\varepsilon = \text{strain} = \Delta L/L. \tag{2.25}$$

In the above equations, A is the area of tube's cross-section, L is the length of the tube and ΔL is the applied displacement boundary condition.

Considering the thickness of the tube's wall to be equal to that of a graphene sheet, $t = 0.34$ nm, the area of the tube, A, will be:

$$A = \pi\left[(R_{CNT} + \frac{t}{2})^2 - (R_{CNT} - \frac{t}{2})^2\right] = \pi D_{CNT} t. \tag{2.26}$$

2.7.2.2 Shear Modulus

In order to obtain the structure's shear modulus, as illustrated in Fig. 2.25, two approaches can be employed, namely, torsion and tensile tests. In the torsion test, the models are twisted by applying an arbitrary angular displacement around their axis and consequently, the corresponding reaction torque is acquired. Finally, the shear modulus of CNTs is calculated by:

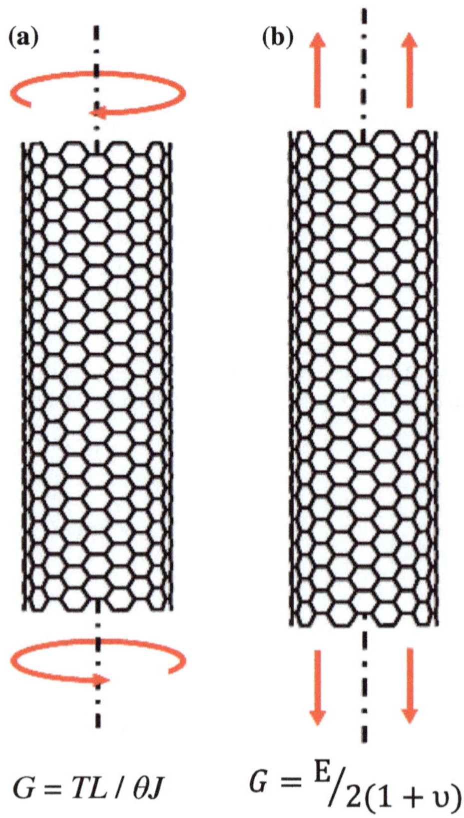

Fig. 2.25 Armchair SWCNT under **a** torsion and **b** tensile test

$$G = TL/\theta J, \tag{2.27}$$

in which θ, T, L and J are the twisting angle, torque, length and polar moment, respectively. As defined below, the polar moment of inertia is:

$$J_n = \sum_{i=1}^{n} J_i, \tag{2.28}$$

where J_{n} is the polar moment and J_{i} is the moment of the ith wall of an n-walled CNT and

$$J_{\mathrm{i}} = \int r^2\, dA = 2\pi \int_{r_{in}}^{r_{out}} r^3.dr = \frac{\pi(r_{out}^4 - r_{in}^4)}{2}. \tag{2.29}$$

For the tensile test, the models are loaded by an arbitrary axial displacement boundary condition and consequently, the reaction force and Poisson's ratio of the models is obtained. Finally, shear modulus of the CNTs is calculated. Comparing the results from tensile and torsion tests and the considerable difference between their results reveals the fact that carbon nanotubes do not behave in an isotropic way and hence, tensile test is not a reliable approach to predict the torsion properties of carbon nanotubes.

2.7.2.3 Poisson's Ratio

Poisson's ratio as stated in Eq. (2.22), is the negative ratio of the circumferential or lateral strain, $\varepsilon_{\mathrm{lateral}}$, to the axial strain, $\varepsilon_{\mathrm{axial}}$. The applied displacement boundary condition results in an elongation of our structures. This elongation causes a reduction in diameter of the whole structure, subsequently. This reaction means that this structure has a positive Poisson's ratio, as it is common for classical engineering materials. Indeed, in some structured materials, mostly polymer foams, Poisson's ratio can be negative which means that by stretching them in one direction, they expand in the perpendicular direction. Therefore, in order to obtain Poisson's ratio, the radial displacement is calculated, as it is shown in Fig. 2.26. However, if Eq. (2.23) of elastic classical mechanics applies, Poisson's ratio can be simply obtained by having Young's modulus and shear modulus values.

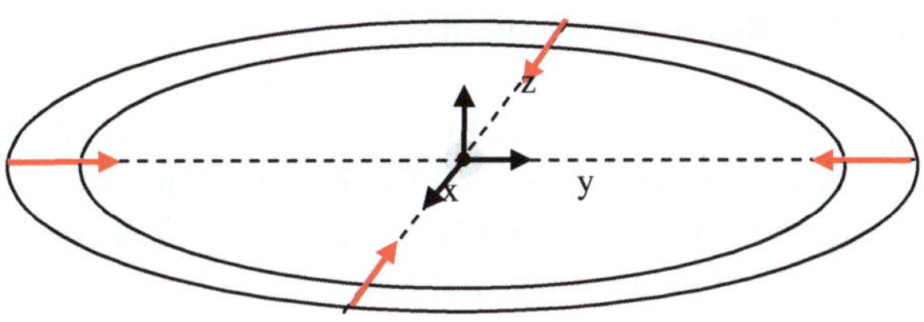

Fig. 2.26 Cross-section of a CNT, showing the lateral forces

2.7.2.4 Buckling Behavior

Buckling is a kind of elastic instability in a structure that occurs under a certain compressive load. In the theory of elasticity, the critical buckling load of a structure, as defined in the Euler equation of buckling load, depends on its geometry, as well as its boundary conditions:

$$P_{cr} = \frac{n^2\pi^2 EI}{(KL)^2}, \tag{2.30}$$

where P_{cr} is the critical buckling load, E is the structure's axial Young's modulus, K is the effective length constant and L is the length of the tube. In the above equation, n defines the buckling mode and I is the structure's second moment of area. As a hollow cylinder is the most similar classical structure to the studied carbon nanotubes, second moment of area defined for a hollow cylinder can be used for obtaining the analytical results, as follows:

$$I = \pi\left[(d+t)^4 - (d-t)^4\right]/64 \tag{2.31}$$

in which t is the thickness of the tube and d is the diameter of the pertaining tube.

The applied boundary conditions play an important role in the reaction of the studied structure under load. This influence is described by the concept of the effective length for a classic column structure that is presented as parameter K in Eq. (2.30). There are four different classical modes for buckling and four pertaining effective length parameters for each. The theoretical K values are defined for a column structure. These different modes of boundary condition and different possible buckling behaviors are shown schematically in Fig. 2.27 and the appropriate theoretical value of K for each case is provided (Rahmandoust and Öchsner 2012a, b).

Let us consider the boundary condition of the carbon nanotubes as fixed from translation and rotation on one end and free to move on the other. As explained before, this boundary condition is known as the Cantilever boundary condition and the corresponding K value for this case is ($K = 2$). For this case, the critical buckling load for both armchair and zigzag structures were modeled by the finite element software MSC.Marc and also it was calculated analytically using Eq. (2.30). Table 2.2 shows the accuracy of the obtained values and Fig. 2.28 illustrates the initial buckling of zigzag and armchair SWCNTs.

2.7.2.5 Resonant Frequency

On the other hand, the natural or resonance frequency of a structure depends on its geometry and mass, as well as applied boundary conditions. For a beam element, under the cantilever boundary conditions, the first mode of resonance frequency in Hz, can be calculated by the following equation:

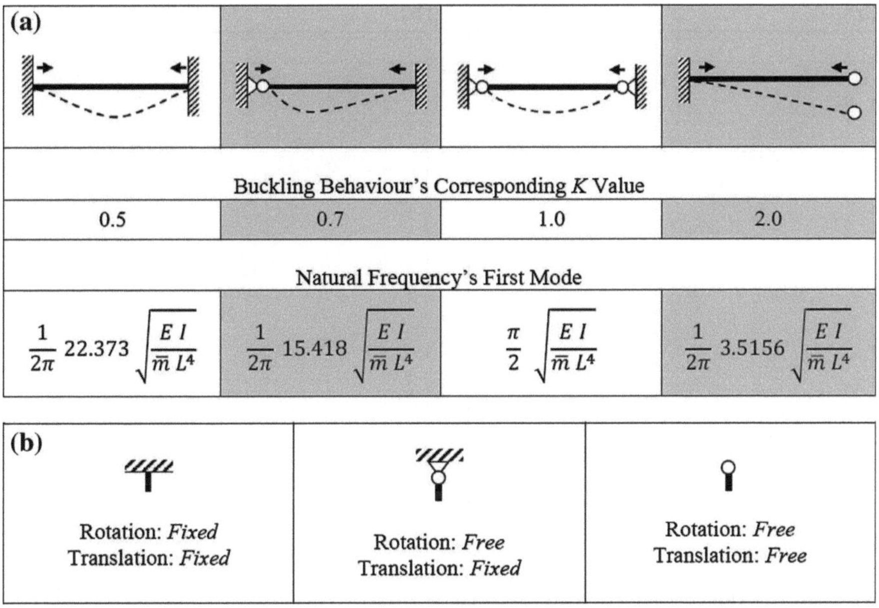

Fig. 2.27 **a** Classical boundary conditions and their corresponding effective length constants for buckling and their first resonance frequency equation, **b** definitions of the symbols of applied boundary conditions

Table 2.2 Comparing the analytical solution of zigzag and armchair SWCNTs critical buckling load, with the optimized finite element results

Carbon nanotube type	Analytical results (nN)	FEM results (nN)	Error (%)
(10, 10) armchair SWCNT	3.668	3.664	0.10
(17, 0) zigzag SWCNT	3.447	3.446	0.03

$$f_1 = \frac{1}{2\pi} 3.5156 \sqrt{\frac{EI}{\bar{m} L^4}}, \tag{2.32}$$

in which $\bar{m}$ is the structure's mass density, i.e. mass per unit length. Second and third modes of frequency are equal to $f_2 = 6.268 \cdot f_1$ and $f_3 = 17.456 \cdot f_1$, respectively (Fan et al. 2009). The FE results for natural frequency of armchair and zigzag SWCNTs were compared to the analytical solution for a hollow cylinder, as shown in Table 2.3.

Like the previous case, the single-walled carbon nanotubes were restricted under a cantilever type boundary condition. By defining the appropriate mass density in the finite element program, the structures' first two modes of natural frequency were simulated. Figure 2.29 shows the different modes of deviation of a (17, 0) single-walled carbon nanotube around its initial straight mode.

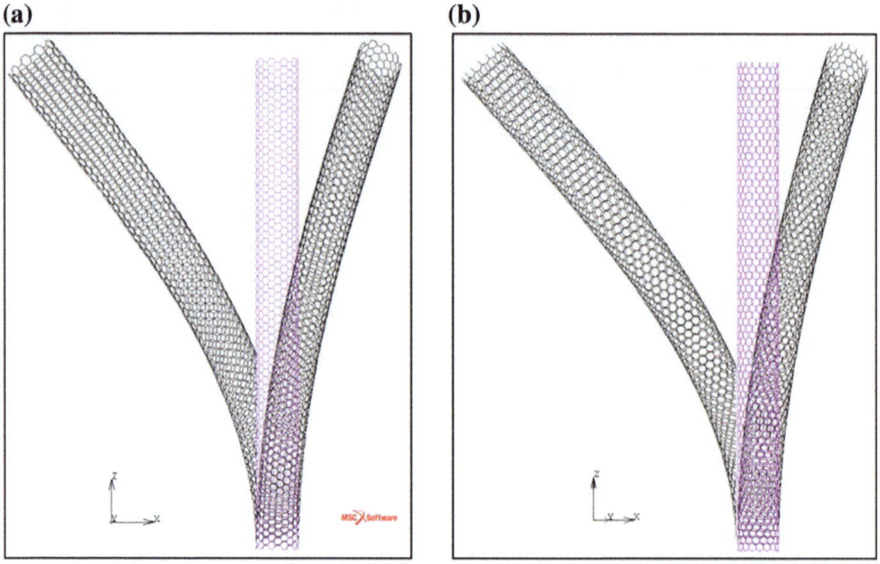

Fig. 2.28 Buckling behavior of **a** armchair and **b** zigzag SWCNTs

Table 2.3 Comparing the analytical solution of zigzag and armchair SWCNTs resonance frequency, with the optimized finite element results

Carbon nanotube type	Results	Frequency mode 1 (GHz)	Frequency mode 2 (GHz)
(10, 10) armchair SWCNT	Analytical	23.97	150.2
	Finite element	23.94	141.1
	Error (%)	0.12	6.05
(17, 0) zigzag SWCNT	Analytical	23.44	146.9
	Finite element	23.43	138.6
	Error (%)	0.04	5.65

2.7.2.6 Single Ring Model

In the modeling of carbon nanotubes, it is necessary to consider a reasonable length for the structure. The length-to-diameter ratio in 1-D nanostructured materials like CNTs is a very important factor. This distinct property is exactly the key of the excellence of nanomaterials compared to other conventional materials. Characterizing the mechanical properties of CNTs by means of numerical simulation, which involves a large number of nodes and elements, is a time consuming process. Hence, being able to obtain valid results by a minimum number of elements will be a valuable advantage in computation-based characterizations of carbon nanotubes.

Hence, the effect of reducing the size of the element from the pristine size down to a single ring was studies. In order to be able to use a single ring of the

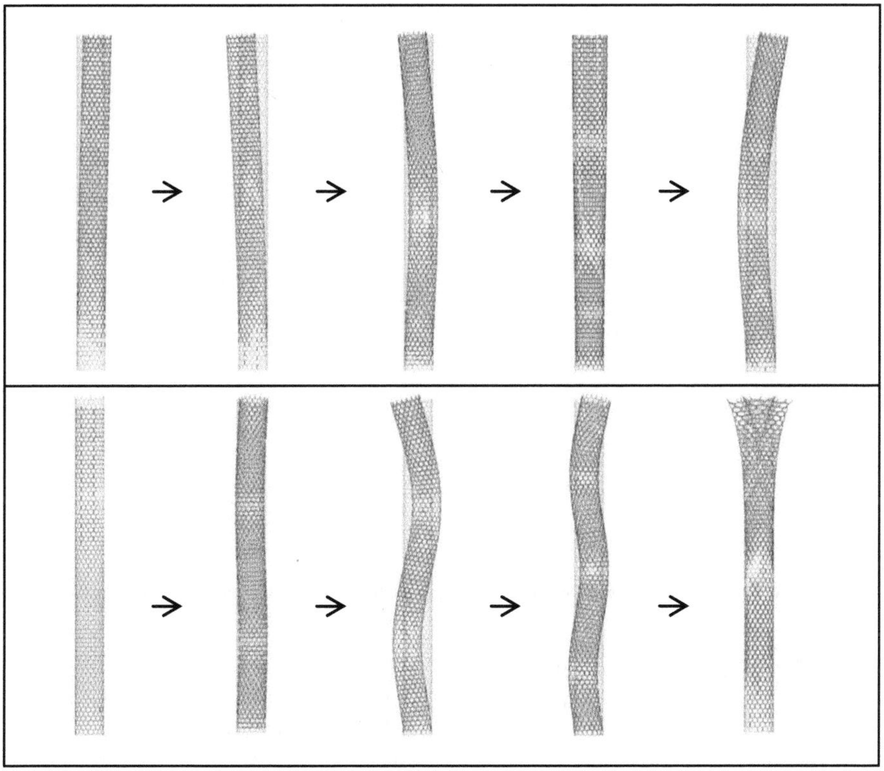

Fig. 2.29 Frequency modes of a zigzag SWCNTs under cantilever boundary condition

SWCNT instead of entire structure, as it is shown schematically in Fig. 2.30, boundary conditions have to be modified in a way that the ring can represent the behavior of the whole structure. For this purpose, the structure was set to be able to move in radial direction but not to rotate tangential to the ring's wall. In addition, translational movement along the ring's main axis was disabled at the lower nodes of the ring (Rahmandoust and Öchsner 2012a, b).

Table 2.4 compares the obtained Young's modulus for a long and a ring SWCNT and gives the error of using a ring instead of the entire tube. As the table shows, the difference between the results, for both zigzag and armchair cases is below half a percent, which proves that the findings are reliable.

2.7.3 Mechanical Properties of MWCNTs

Similar to SWCNTs major mechanical properties of MWCNTs can also be achieved using FE modeling. In case of DWCNTs, different approximations were tested to see how these approaches influences the accuracy of the obtained results. First, the

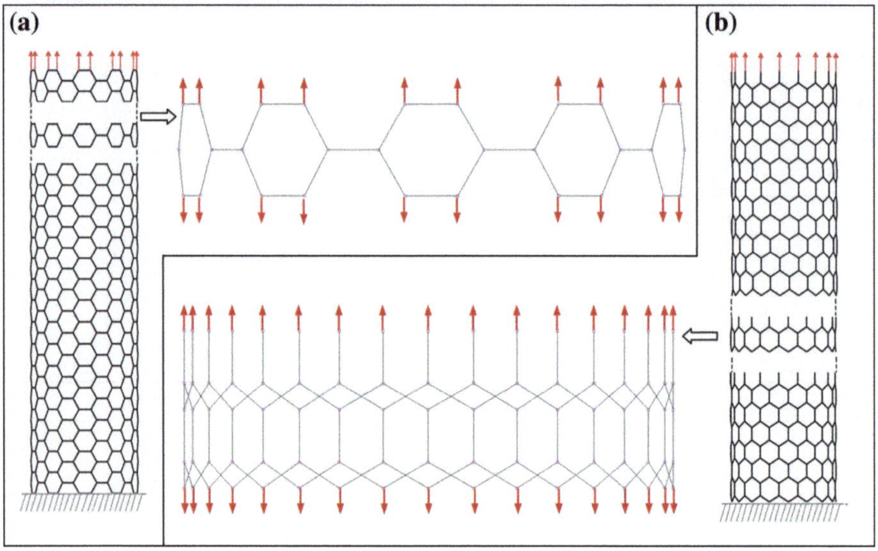

Fig. 2.30 Schematic view of the long and the ring structures for **a** armchair and **b** zigzag SWCNTs

Table 2.4 Young's modulus of a long and ring SWCNTs

	Perfect form; pristine tube		Modified boundary conditions; single ring		Percentage of error	
SWCNT type	Armchair	Zigzag	Armchair	Zigzag	Armchair	Zigzag
E (TPa)	1.042	1.034	1.047	1.038	0.48 %	0.39 %
Number of nodes	12,280	12,036	70	102	–	–
Length (nm)	50.307	50.268	0.246	0.426	–	–

Lennard-Jones force equation was replaced with a 5th degree polynomial approximation; then, this accurate approximation of the Lennard-Jones force was totally ignored and the obtained results were compared. Finally, as it is going to be explained later, an equivalent spring constant was defined in a way that to approximate the Lennard-Jones force, in its effective distance only, with a linear function of displacement. Furthermore, the DWCNTs were generated as a single ring and a long tube, with and without defining the Lennard-Jones interaction. Tables 2.5 and 2.6 displays the effect of the tested approximations on Young's modulus and shear modulus of DWCNTs, respectively.

Based on the explained approximations, single- to 5-walled carbon nanotube models were generated and analyzed in the finite element software. These CNTs and their characteristics are presented in Table 2.7. Figure 2.31 shows the Young's modulus versus the number of walls, in a multi-walled carbon nanotube, where the thickness of the tube's wall was assumed to be equal to 0.34 nm in all models. The non-linear converging trend of E in the diagram implies that by increasing the number of walls, the effect of the inner layers of carbon over the whole

Table 2.5 Young's modulus of DWCNT models in form of a ring and a long tube, with and without the Lennard-Jones force being defined

Type of DWCNT		Young's modulus E (TPa)	
		Long tube; perfect form	Single ring; modified boundary conditions
Armchair	With Lennard-Jones	1.053	1.059
	Without Lennard-Jones	1.051	1.056
Zigzag	With Lennard-Jones	1.033	1.044
	Without Lennard-Jones	1.037	1.045

Table 2.6 Shear modulus of DWCNT models in form of a ring and a long tube, with and without the Lennard-Jones force being defined

Type of DWCNT		Shear modulus G (TPa)	
		Long tube; perfect form	Single ring; modified boundary conditions
Armchair	With Lennard-Jones	0.448	0.514
	Without Lennard-Jones	0.488	0.511
Zigzag	With Lennard-Jones	0.424	0.499
	Without Lennard-Jones	0.470	0.494

Table 2.7 Simulated CNTs and their characteristics

CNT	CNT type	Chirality (m,n)	Length (nm)	Outer radius (nm)
SWCNT	Armchair	(10,10)	15.74	0.6783
	Zigzag	(14,0)	15.62	0.5482
DWCNT	Armchair	(10,10)-(15,15)	15.74	1.0175
	Zigzag	(14,0)-(23,0)	15.62	0.9007
3-WCNT	Armchair	(10,10)-(15,15)-(20,20)	15.74	1.3566
	Zigzag	(14,0)-(23,0)-(32,0)	15.62	1.2532
4-WCNT	Armchair	(10,10)-(15,15)-(20,20)-(25,25)	15.74	1.6958
	Zigzag	(14,0)-(23,0)-(32,0)-(41,0)	15.62	1.6057
5-WCNT	Armchair	(10,10)-(15,15)-(20,20)-(25,25)-(30,30)	15.74	2.0350
	Zigzag	(14,0)-(23,0)-(32,0)-(41,0)-(50,0)	15.62	1.9582

structure reduces toward reaching a rather constant maximum Young's modulus for multi-walled carbon nanotubes.

2.8 Modeling of Defects

Defects are inevitable parts of most of engineering material. In some cases, these defects are purposely introduced to the structure to achieve specifically desired properties and in some other cases, defects occur as a result of the nanostructure's

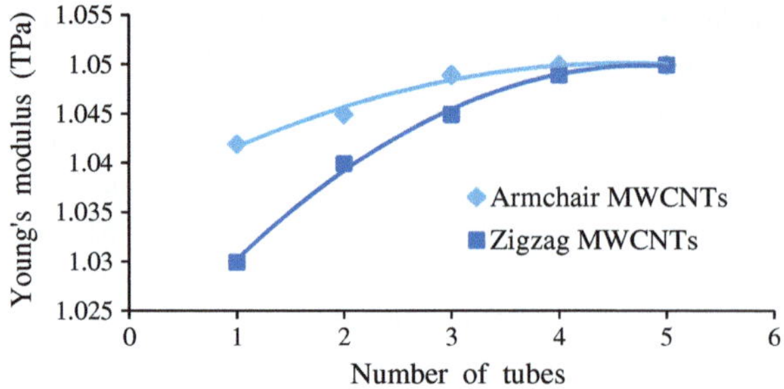

Fig. 2.31 Young's modulus versus the number of walls in zigzag and armchair MWCNTs

fabrication process. However, either case to be the reason, it is necessary to discover the influence of the defect to the resulted properties of the structure.

2.8.1 Perturbation in Carbon Nanotube's Structure

Studying the effect of misplacement of atoms in carbon nanotube's hexagonal lattice or perturbations, as shown schematically in Fig. 2.32, is very interesting, due to being very likely to happen in experimental studies. When carbon nanotubes are produced experimentally, as a result of local stresses and the effect of environmental pressures, misplacement of atoms are unavoidable. Thus, it is important to control the effect of this defect on the properties of the structure. In order to simulate this type of imperfection, the location of atoms were allowed to change randomly by a specific percent of their distance with respect to each other.

Therefore, the effect of 5–15 % of perturbation in the structure was studied on the mechanical properties, as shown in Fig. 2.33. By introducing perturbation to the structures of both zigzag and armchair single-walled carbon

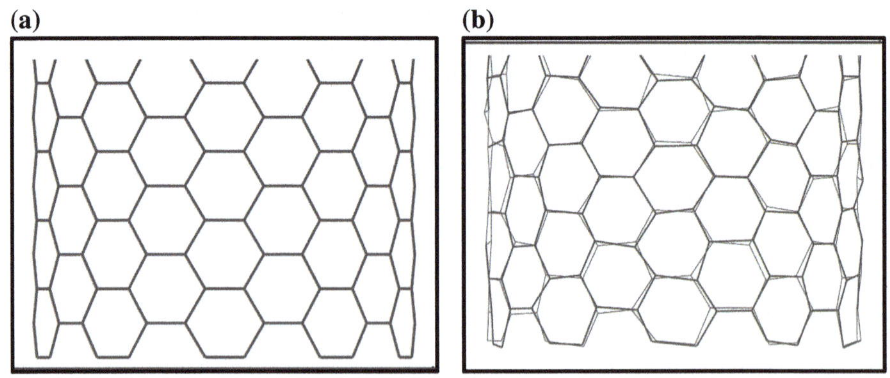

Fig. 2.32 Front view of the **a** perfect and **b** 10 % perturbated SWCNT

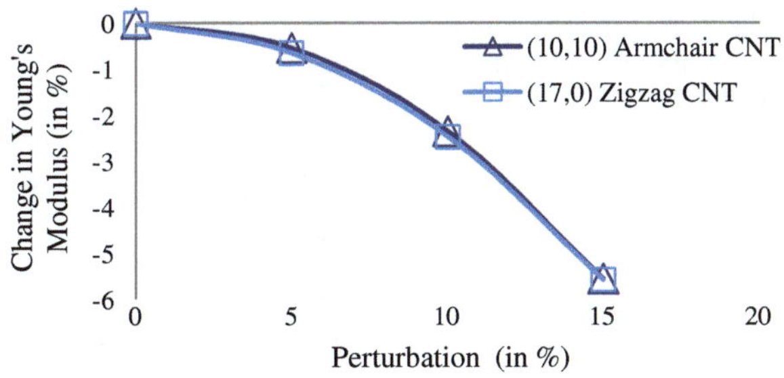

Fig. 2.33 Perturbation versus Young's modulus

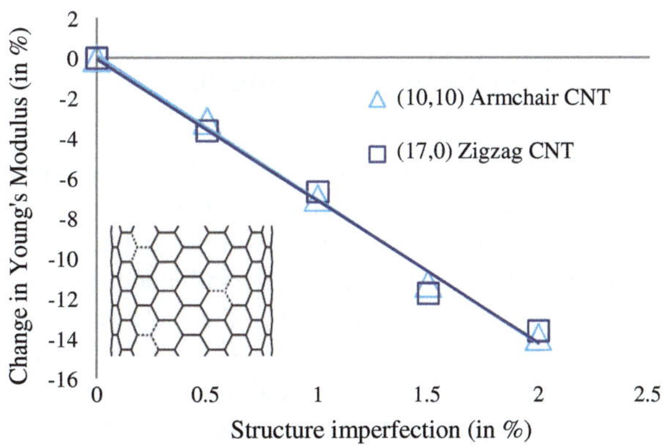

Fig. 2.34 Missing atoms versus Young's modulus

nanotubes, dropdown in Young's modulus was observed. This reduction is not very remarkable in the first 5 % of perturbation but when the structures deviate more from the perfect arrangement, the change in the Young's modulus becomes noticeable (Fig. 2.34).

2.8.2 *Atom-Vacancies in CNT Structure*

Modeling the effect of some missing atoms on Young's modulus of the two main types of SWCNTs has also been investigated to show that as a result of this defect, by an increase in the number of missing carbons, the structures lose their stiffness rather linearly and very fast, as illustrated in Fig. 2.35. For example, the structure loses 10 % of its Young's modulus when only 1.3 % of the atoms are lost. From the experimental point of view, it is inevitable that the structure changes the arrangement of its constituent carbons from hexagonal to pentagonal or heptagonal forms, in order to minimize its internal potential energy, leading to creation of Stone–Wales defect as explained before. Hence, local strains happen in these locations and the Young's modulus is affected consequently.

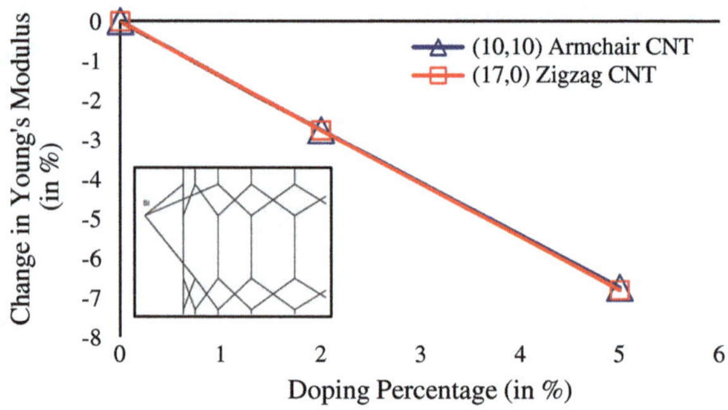

Fig. 2.35 Effect Si doping on Young's modulus of SWCNTs

2.8.3 Doped Carbon Nanotubes

This defect is one of the most likely imperfections happening when a definite number of carbon atoms in the perfect structure of the CNT are replaced with other atoms, which causes considerable undesirable changes in the mechanical properties, but perhaps desirable influence on the electrical or chemical properties of the CNT structure. Other than silicon, which is explained here, any other element can also be introduced to the structure to reveal various changes in the mechanical, chemical and electrical properties of the nanostructure, depending on the type of the doped atom and the number of its valence electrons; proper geometrical and material properties should be assigned in the model for the introduced covalent bond. For random dispersion of the imperfections into the structure of CNTs, Matlab programming was performed and results were imposed to the finite element model (Rahmandoust and Öchsner 2009).

The geometric properties for C–C and Si–C bonds are assumed to be nearly the same because the difference in the covalent radius of carbon and silicon is only 28 pm (Cordero et al. 2008) but as a beam element, their material properties are noticeably different. For example $E_{\text{Si-C}} = 7.18327 \times 10^{-6}$ N/nm^2 (Song et al. 2006), whereas $E_{\text{C-C}} = 5.484 \times 10^{-6}$ N/nm^2 (Rahmandoust and Öchsner 2009).

For the simulation of Si–C bonds, some percentage of the total number of carbon atoms was randomly chosen in the structure of CNT to be replaced with silicon. Since each carbon atom is encircled by three other neighbor atoms, the centre atom emerges radially out and some hillocks are formed on the structure (Lu 1997), so that the length of the connecting beam elements equal to that of the Si–C bond, as it is schematically shown inside Fig. 2.35. Furthermore, as the diagram implies, although the Young's modulus of Si–C bond itself is greater than C–C bond's E, doping Si atoms into the structure of zigzag and armchair CNTs causes a dramatic reduction in the value of CNT's Young's modulus. The reason is that by increasing the percentage of doped atoms, more hillocks are formed in the structure and Young's modulus of carbon nanotube decreases consequently.

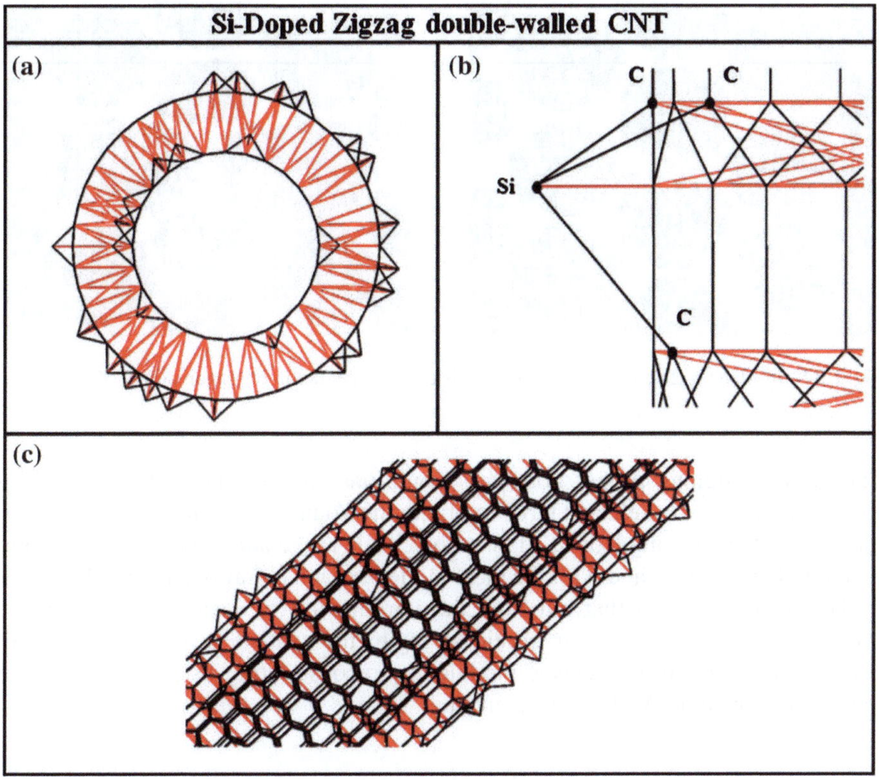

Fig. 2.36 **a** Top, **b** zoomed and **c** global view of a Si-doped DWCNT

Similar to SWCNTs, Si-doped MWCTs show an almost linear decrease in the value of shear modulus by increasing the percentage of Si-doping in CNT models. Figure 2.36 shows a Si-doped DWCNT.

2.8.4 Stone–Wales Defect

The Stone–Wales defect, as introduced earlier in this chapter, is the most abundantly detected defect in the structure of carbon nanotubes that occurs naturally after initial formation of vacancies (Charlier 2002). A Stone–Wales defect can be imagined to be formed by a way of an in–plane $\pi/2$ rotation of a C–C bond (Pozrikidis 2009), as schematically illustrated in Fig. 2.37.

Therefore, for the modeling of the defect, the easiest way was to create the model first in planar form and then to roll it into a tube after rotating the randomly selected bonds for $\pi/2$ degree in-plane. The models with the Stone–Wales defect should be finally optimized by minimizing the Tersoff-Brenner potential.

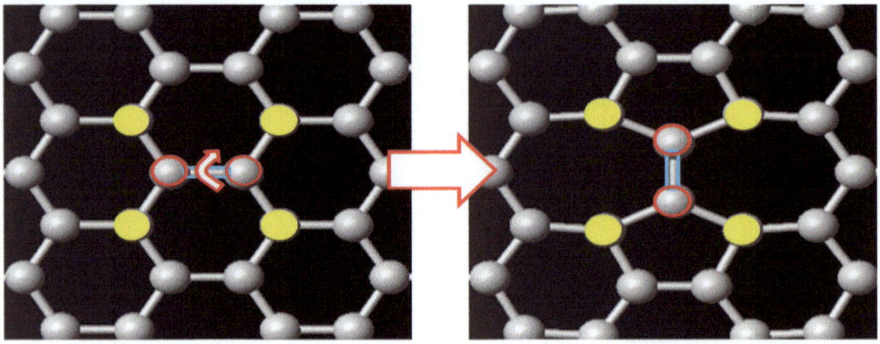

Fig. 2.37 In-plane π/2 degree rotation of bonds to make pentagon-heptagon defect

Then after assigning proper boundary conditions, an external compression load can be applied to the defected CNT to obtain the Young's modulus, E, of the structure was obtained using Hooke's law in the elastic range and the results were compared with each other and with the ideal mode for each chirality and length.

The results showed that although the introduction of vacancy defects to the body of CNTs weakens their elastic strength, the system is almost capable of restoring its original high elastic modulus by reconstructing its atoms in form of the 5-7-7-5 or Stone–Wales arrangement.

2.8.5 Hetero-Junction CNTs

A local view on the structure of hetero-junction CNTs shows that these nanostructures are made up of the same carbon hexagonal unit blocks, as perfect CNTs. However, the wider view reveals that hetero-junction CNTs are constructed of two CNTs with different chiralities, connected together by a kink, having pentagon-heptagon defect, as shown in Fig. 2.38. These kinks, whose existence and size seem to be noticeably effective on the mechanical properties of hetero-junction

Fig. 2.38 Occurrence of pentagon-heptagon defect on a kink

CNTs, are generally divided into two groups, namely, straight and bent connections (Ghavamian et al. 2015).

For the finite element simulation of the hetero-junction CNTs, CoNTub software can be employed, which is an algorithm for connecting two arbitrary carbon nanotubes. However, since the structure of hetero-junction CNTs comprises of two tubes and a kink part, the structure does not make a homogeneous body. Hence, analytical calculations to determine the moduli and critical buckling loads of hetero-junction CNTs required to be modified with a proper approximation. According to this approximation, each hetero-junction CNT was assumed to be a homogeneous CNT with a constant area, A_{Hetero} and a constant diameter, D_{Hetero}, using the general weight percentage relation, as follows:

$$L_{\mathrm{Hetero}} = L_{\mathrm{Tube1}} + L_{\mathrm{Kink}} + L_{\mathrm{Tube2}}; \tag{2.33}$$

where

$$L_{\mathrm{Kink}} = \frac{\sqrt{3}}{2}\pi(D_d), \tag{2.34}$$

and

$$D_{\mathrm{d}} = d_{\text{wider tube}} - d_{\text{thinner tube.}} \tag{2.35}$$

Hence,

$$A_{Hetero} = \frac{L_{wider-tube} \times A_{wider-tube} + L_{Kink} \times A_{Kink} + L_{Thinner-tube} \times A_{Thinner-tube}}{L_{Hetero}} \tag{2.36}$$

$$D_{Hetero} = \frac{L_{wider-tube} \times D_{wider-tube} + L_{Kink} \times D_{Kink} + L_{Thinner-tube} \times D_{Thinner-tube}}{L_{Hetero}} \tag{2.37}$$

and

$$A_{Kink} = \frac{A_{wider-tube} + A_{Thinner-tube}}{2}. \tag{2.38}$$

References

Ahmadi, M.T., Webb, J.F., Razali, I., Rahmandoust, M.: Carbon-based materials concepts and basic physics. In: Razali, I., Ahmadi, M.T., Anwar, S. (eds.) Advanced Nanoelectronics, pp. 49–82. CRC Press, Boca Raton (2012)

Ajayan, P.M., Ravikumar, V., Charlier, J.-C.: Surface reconstruction and dimensional changes in single-walled carbon nanotubes. Phys. Rev. Lett. **81**, 1437–1440 (1998)

Azadi, S., Moradian, R., Shafaee, A.M.: The effect of Stone-Wales defect orientations on the electronic properties. Comp. Mater. Sci. **49**, 699–703 (2010)

Baierle, R.J., et al.: Electronic and structural properties of silicon-doped carbon nanotubes. Phys. Rev. B **64**, 085413-1–085413-4 (2001)

Berber, S., Kwon, Y.-K., Tomànek, D.: Unusually high thermal conductivity of carbon nanotubes. Phys. Rev. Lett. **84**(20), 4613–4616 (2000)

Bethune, D.S., et al.: Cobalt–catalysed growth of carbon nanotubes with single–atomic–layer walls. Nature **363**, 605–607 (1993)

Blase, X., et al.: Boron-mediated growth of long helicity-selected carbon nanotubes. Phys. Rev. Lett. **83**, 5078–5081 (1999)

Carroll, D.L., et al.: Effects of nanodomain formation on the electronic structure of doped carbon nanotubes. Phys. Rev. Lett. **81**, 2332–2335 (1998)

Chang, T., Gao, H.: Size-dependent elastic properties of single-walled carbon nanotubes via a molecular mechanics model. J. Mech. Phys. Solids **51**, 1059–1074 (2003)

Charlier, J.-C.: Defects in carbon nanotubes. Acc. Chem. Res. **35**, 1063–1069 (2002)

Chico, L., et al.: Pure carbon nanoscale devices: nanotube heterojunctions. Phys. Rev. Lett. **76**, 971–974 (1996)

Choi, J.H., Ihm, J., Louie, S.G., Cohen, M.L.: Defects, quasibound states, and quantum conductance in metallic carbon nanotubes. Phys. Rev. Lett. **84**, 2917–2920 (2000)

Collins, P.G.: Defects and disorder in carbon nanotubes. In: Narlikar, A.V., Fu, Y.Y. (eds.) Oxford Handbook of Nanoscience and Technology: Frontiers and Advances. Oxford University Press, Oxford (2009)

Cordero, B., et al.: Covalent radii revisited. Dalton Trans. **21**, 2832–2838 (2008)

Czerw, R., et al.: Identification of electron donor states in N-doped carbon nanotubes. Nano Lett. **1**(9), 457–460 (2001)

Deck, C.P., Vecchio, K.: Growth mechanism of vapor phase CVD-grown multi-walled carbon nanotubes. Carbon **43**(12), 2608–2617 (2005)

DeHon, A.: Array-based architecture for molecular electronics. In: First Workshop on Non-Silicon Computation (2002)

Dekker, C.: Carbon nanotubes as molecular quantum wires. Phys. Today **52**(5), 22–28 (1999)

Dresselhaus, M.S., Dresselhaus, G., Saito, R.: Physics of carbon nanotubes. Carbon **33**(7), 883–891 (1995)

Dresselhaus, M.S., Dresselhaus, G., Charlier, J.C., Hernandez, E.: Electronic, thermal and mechanicalproperties of carbon nanotubes. Phil. Trans. R. Soc. Lond. A **362**, 2065–2098 (2004)

Dresselhaus, M.S., Dresselhaus, G., Saito, R., Jorio, A.: Raman spectroscopy of carbon nanotubes. Phys. Rep. **409**(2), 47–99 (2005)

Duan, W.H., Wang, Q., Liew, K.M., He, X.Q.: Molecular mechanics modeling of carbon nanotube fracture. Carbon **45**, 1769–1776 (2007)

Dürkop, T., Getty, A.S., Cobas, E., Fuhrer, M.S.: Extraordinary mobility in semiconducting carbon nanotubes. Nano Lett. **4**, 35–39 (2004)

Ebbesen, T.W., Ajayan, P.M.: Large scale synthesis of carbon nanotubes. Nature **358**, 220–222 (1992)

Ehrhart, P.: Properties and interactions of atomic defects in metals and alloys. In: Landolt-Börnstein, New Series III, p. 88. Springer-Verlag, Berlin (1991)

Fagan, S.B., et al.: Ab initio study of an organic molecule interacting with a silicon-doped carbon nanotube. Diam. Relat. Mater. **12**, 861–863 (2003)

Fakhrabadi, M.M.S., Rastgoo, A., Ahmadian, M.T.: Pull-in behaviors of carbon nanotubes with vacancy defects and residual stresses. J. Comput. Theor. Nanosci. **11**(1), 153–159 (2014)

Fan, C.W., Liu, Y.Y., Hwu, C.: Finite element simulation for estimating the mechanical properties of multi-walled carbon nanotubes. Appl. Phys. A **95**, 819–831 (2009)

Fang, S.C., Chang, W.J., Wang, Y.H.: Computation of chirality- and size-dependent surface Young's moduli for single-walled carbon nanotubes. Phys. Rev. Lett. A **371**, 499–503 (2007)

Farsadi, M., Öchsner, A., Rahmandoust, M.: Numerical investigation of composite materials reinforced with waved carbon nanotubes. J. Compos. Mater. **47**, 1425–1434 (2013)

Fereidoon, A., Rafiee, R., Moghadam, R.M.: A modal analysis of carbon-nanotube-reinforced polymer by using a multiscale finite-element method. Mech. Compos. Mater. **49**(3), 325–332 (2013)

Formica, G., Lacarbonara, W., Alessi, R.: Vibrations of carbon nanotube-reinforced composites. J. Sound Vibrat. **329**, 1875–1889 (2010)

Gai, P., et al.: Structural systematics in boron-doped single wall carbon nanotubes. J. Mater. Chem. **14**, 669 (2004)

Gal'Pern, E.G., Stankevich, I.V., Chistyakov, A.L., Chernozatonskiĭ, L.A.: Atomic and electronic structure of the barrelenes b-Cm with m = 36 + 12n. JETP Lett. **55**, 483–486 (1992)

Gao, R.P., et al.: Nanomechanics of individual carbon nanotubes from pyrolytically grown arrays. Phys. Rev. Lett. **85**, 622–625 (2000)

Ge, M., Sattler, K.: Bundles of carbon nanotubes generated by vapor-phase growth. Appl. Phys. Lett. **64**(6), 710–711 (1994)

Ghadyani, G., Rahmandoust, M.: Computational nanomechanics investigation techniques. In: Silvestre, N. (ed.) Advanced Computational Nanomechanics. Wiley, New York (2015)

Ghavamian, A., Öchsner, A.: Numerical investigation on the influence of defects on the buckling behavior of single-and multi-walled carbon nanotubes. Physica E **46**, 241–249 (2012)

Ghavamian, A., Öchsner, A.: Numerical modeling of eigenmodes and eigenfrequencies of single- and multi-walled carbon nanotubes under the influence of atomic defects. J. Nano Res-SW **21**, 158–164 (2013a)

Ghavamian, A., Öchsner, A.: Numerical modeling of eigenmodes and eigenfrequencies of single- and multi-walled carbon nanotubes under the influence of atomic defects. Comp. Mater. Sci. **72**, 42–48 (2013b)

Ghavamian, A., Rahmandoust, M., Öchsner, A.: A numerical evaluation of the influence of defects on the elastic modulus of single and multi-walled carbon nanotubes. Comp. Mater. Sci. **62**, 110–116 (2012)

Ghavamian, A., Rahmandoust, M., Öchsner, A.: On the determination of the shear modulus of carbon nanotubes. Compos. Part B-Eng. **44**, 52–59 (2013)

Ghavamian, A., Rahmandoust, M., Öchsner, A.: Perfect and Defective Hetero-Junction CNTs. In: Silvestre, N. (ed.) Advanced Computational Nanomechanics. Wiley, New York (2015) in-press

Guo, T., et al.: Catalytic growth of single-walled nanotubes by laser vaporization. Chem. Phys. Lett. **243**, 49–54 (1995)

Hernández, E., Goze, C., Bernier, P., Rubio, A.: Elastic properties of single-wall nanotubes. Appl. Phys. A **68**, 287–292 (1999a)

Hernández, E., Goze, C., Bernier, P., Rubio, A.: Elastic properties of C and BxCyNz composite nanotubes. Phys. Rev. Lett. **80**, 4502 (1999b)

Hosseini, A.A., Allahyari, M., Besheli, S.D.: Synthesis of carbon nanotubes, nano fibbers and nano union by electric arc discharge method using nacl accuse as solution and fe and ni particles as catalysts. Int. J. Sci. Env. Technol. **1**(3), 217–229 (2012)

Hsu, W., et al.: Boron doping effects in carbon nanotubes. J. Mater. Chem. **10**, 1425–1429 (2000)

Unidym-Technology: CNT manufacturing. http://www.unidym.com/technology/cnt_manufacture.html (2008)

Iijima, S.: Helical microtubules of graphitic carbon. Nature **354**, 56–58 (1991)

Iijima, S., Ichihashi, T.: Single-shell carbon nanotubes of 1-nm diameter. Nature **363**, 603–605 (1993)

Jahanshahi, M., Kiadehi, A.D.: Fabrication, purification and characterization of carbon nanotubes: arc-discharge in liquid media (ADLM). In: Suzuki, S. (ed.) Syntheses and Applications of Carbon Nanotubes and Their Composites, pp. 55–76. Croatia, InTech (2013)

Javey, A., et al.: Ballistic carbon nanotube transistors. Nature **424**, 654–657 (2003)

Jin, Y., Yuan, F.G.: Simulation of elastic properties of single–walled carbon nanotubes. Compos. Sci. Technol. **63**, 1507–1515 (2003)

Jorio, A., Dresselhaus, G., Dresselhaus, M.S.: Carbon Nanotubes: Advanced Topics in the Synthesis, Structure, Properties and Applications. Springer, Heidelberg (2008)

Journet, C., et al.: Large scale production of single walled carbon nanotubes by the electric arc technique. Nature **388**, 756–758 (1997)

Kalamkarov, A.L., et al.: Analytical and numerical techniques to predict carbonnanotube properties. Int. J. Solids Struct. **43**, 6832–6854 (2006)

Kaptay, G., Sytchev, J.: Report. University of Miskolc (2005)
Kataura, H. et al.: Optical properties of single-wall carbon nanotubes. In: International Conference on Science and Technology of Synthetic Metals (1999)
Kaw, A.K.: Mechanics of composite materials, 2nd edn. CRC Press Taylor and Francis, Boca Raton (2006)
Kelly, M.J.: Low-dimensional semiconductors: materials, physics, technology, devices. Oxford University, Oxford (1995)
Keskar, G., et al.: Nitrogen doping and characterization of isolated single-wall carbon nanotubes using liquid precursors. Chem. Phys. Lett. **412**, 269–273 (2005)
Kivistö, S., et al.: Carbon nanotube films for ultrafast broadband technology. Opt. Express **17**, 2358–2363 (2009)
Kosakovskaya, Z.Y., Chernozatonskii, L.A., Fedorov, E.A.: Nanofilament carbon structure. JETP Lett. **56**, 26–29 (1992)
Krishnan, A., et al.: Young's modulus of single-walled nanotubes. Phys. Rev. B **58**, 14013–14019 (1998)
Li, C., Chou, T.-W.: A structural mechanics approach for the analysis of carbon nanotubes. Int. J. Solids Struct. **40**(10), 2487–2499 (2003)
Li, W.Z., et al.: Large scale synthesis of aligned carbon nanotubes. Science **274**(5293), 1701–1703 (1996)
Liu, J. et al.: Chirality-controlled synthesis of single-wall carbon nanotubes using vapour-phase epitaxy. Nat. Commun. **3**, 1199-1-7 (2012)
Liu, X. et al.: Detailed analysis of the mean diameter and diameter distribution of single-wall carbon nanotubes from their optical response. Phys. Rev. B **66**, 4 (2002)
Lu, J.P.: Elastic properties of carbon nanotubes and nanoropes. Phys. Rev. Lett. **79**, 1297–1300 (1997)
Lu, A.J., Pan, B.C.: Nature of single vacancy in achiral carbon nanotubes. Phys. Rev. Lett. **92**, 105504 (2004)
Margulis, A.V.: Theoretical estimations of third-order optical nonlinearities for semiconductor carbon nanotubes. J. Phys.: Condens. Matter **11**, 3065 (1999)
Margulis, A.V., Sizikova, T.A.: Theoretical study of third order nonlinear optical response of semiconductor carbon nanotube. Phys. B **245**, 173–189 (1998)
Martinez, A., Yamashita, S.: Carbon nanotube-based photonic devices: applications in nonlinear optics. In: Marulanda, J.M. (ed.) Carbon Nanotubes Applications on Electron Devices. Croatia, InTech (2011)
McEuen, P.L.: Nanotechnology: carbon-based electronics. Nature **393**, 15–17 (1998)
McGuire, K., et al.: Synthesis and Raman characterization of boron-doped single-walled carbon nanotubes. Carbon **43**, 219–227 (2005)
Moghadam, R.M., Hosseini, S.A., Salehi, M.: The influence of Stone–Thrower–Wales defect on vibrational characteristics of single-walled carbon nanotubes incorporating Timoshenko beam element. Physica E **62**, 80–89 (2014)
Nardelli, M.B., Yakobson, B.I., Bernholc, J.: Brittle and ductile behavior in carbon nanotubes. Phys. Rev. Lett. **81**, 4656–4659 (1998)
NASA: NASA's Goddard Space Flight Center Report. National Aeronautics and Space Administration (NASA) (2005)
Natsuki, T., Tantrakarn, K., Endo, M.: Prediction of elastic properties for single–walled carbon nanotubes. Carbon **42**, 39–45 (2004)
Nikolaev, P., et al.: Gas phase catalytic growth of single-walled carbon nanotubes from carbon monoxide. Chem. Phys. Lett. **313**, 91–97 (1999)
Ponomareva, I., Chernozatonskii, L.A., Andriotis, A.N., Menon, M.: Formation pathways for single-wall carbon nanotube multiterminal junctions. New J. Phys. **5**, 119.1–12 (2003)
Pop, E., et al.: Thermal conductance of an individual single-wall carbon nanotube above room temperature. Nano Lett. **6**(1), 96–100 (2006)
Pozrikidis, C.: Effect of the Stone-Wales defect on the structure and mechanical properties of single-wall carbon nanotubes in axial stretch and twist. Arch. Appl. Mech. **79**, 113–123 (2009)

Pullen, A., Zhao, G.L., Bagayoko, D., Yang, L.: Structural, elastic, and electronic properties of deformed carbon nanotubes under uniaxial strain. Phys. Rev. B **71**, 205410–205415 (2005)

Rafique, M.M.A., Iqbal, J.: Production of carbon nanotubes by different routes—a review. J. Encapsul. Adsorpt. Sci. **1**, 29–34 (2011)

Rahmandoust, M., Öchsner, A.: Young's modulus variation of carbon nanotubes due to defects associated with atomic reconstruction of random vacancies. J. Comput. Theor. Nanosci. **12**, (2015) in press

Rahmandoust, M., Öchsner, A.: Influence of structural imperfections and doping on the mechanical properties of single-walled carbon nanotubes. J. Nano Res-SW **6**, 185–196 (2009)

Rahmandoust, M., Öchsner, A.: Buckling behaviour and natural frequency of zigzag and armchair single-walled carbon nanotubes. J. Nano Res-SW **16**, 153–160 (2012a)

Rahmandoust, M., Öchsner, A.: On finite element modeling of single and multi-walled carbon nanotubes. J. Nanosci. Nanotechnol. **12**, 8129–8136 (2012b)

Ray, M.C., Kundalwal, S.I.: Effect of carbon nanotube waviness on the load transfer characteristics of short fuzzy fiber-reinforced composite. J. Nanomech. Micromech. **4**, A4013010 (2013)

Reich, S., Thomsen, C., Maultzsch, J.: Carbon Nanotubes: Basic Concepts and Physical Properties. Wiley-VCH, Weinheim (2004)

Resasco, D.E., et al.: A scalable process for production of single-walled carbon nanotubes (SWNTS) by catalytic disproportionation of Co on a solid catalyst. J. Nanopart. Res. **4**, 131–136 (2002)

Ruoff, R.S., Lorents, D.C.: Mechanical and thermal properties of carbon nanotubes. Carbon **33**(7), 925–930 (1995)

Saito, R., Fujita, M., Dresselhaus, G., Dresselhaus, M.S.: Electronic structure of chiral graphene tubules. Appl. Phys. Lett. **60**, 2204–2206 (1992)

Saito, R., Dresselhaus, M.S., Dresselhaus, G.: Tunneling conductance of connected carbon nanotubes. Phys. Rev. B **53**, 2044–2049 (1996)

Saito, R., Dresselhaus, G., Dresselhaus, M.S.: Physical Properties of Carbon Nanotubes. Imperial College Press, London (1998)

Samsonidze, G.G., et al.: Family behavior of the optical transition energies in single-wall carbon nanotubes of smaller diameters. Appl. Phys. Lett. **85**(23), 5703–5705 (2004)

Sánchez-Portal, D., et al.: Ab initio structural, elastic, and vibrational properties of carbon nanotubes. Phys. Rev. B **59**, 12678–12688 (1999)

Santos, H.D.L.: Introduction to microelectromechanical microwave systems. Artech House Publishers, London (1999)

Scarselli, M., Castrucci, P., Crescenzi, M.D.: Electronic and optoelectronic nano-devices based on carbon nanotubes. J. Phys. **24**, 313202-1-36 (2012)

Sen, R., et al.: Nitrogen-containing carbon nanotubes. J. Mater. Chem. **7**, 2335–2337 (1997)

Sie, C.T.: Carbon nanotube reinforced composites: metal and ceramic matrices. Wiley–VCH, Weinheim (2009)

Sinha, S., et al.: Off-axis thermal properties of carbon nanotube films. J. Nanopart. Res. **7**(6), 651–657 (2005)

Song, H.Y., Sun, H.M., Zhang, G.X.: Molecular dynamic study of effects of Si-doping upon structure and mechanical properties of carbon nanotubes. Commun. Theor. Phys. **45**, 741–744 (2006)

Song, X., Ge, Q., Yen, S.-C.: A first-principles study on the elastic properties of single-walled carbon nanotubes. J. Nanoeng. Nanosys. **223**, 163–168 (2010)

Stone, A.J., Wales, D.J.: Theoretical studies of icosahedral C60 and some related structures. Chem. Phys. Lett. **128**, 501–503 (1986)

Tanaka, K., et al.: Interlayer interaction of two graphene sheets as a model of double-layer carbon nanotubes. Carbon **35**, 121–125 (1997)

Tang, Z.K., et al.: Superconductivity in 4 angstrom single-walled carbon nanotubes. Science **292**(5526), 2462–2465 (2001)

Tans, S.J., Verschueren, M., Dekker, C.: Room-temperature transistor based on a single carbon nanotube. Nature **393**, 49–51 (1998)

Terrones, M. et al.: Molecular junctions by joining single-walled carbon nanotubes. Phys. Rev. Lett. **89**(7), 075505-1-4 (2002)

Terrones, M., et al.: Carbon nitride nanocomposites: formation of aligned CxNy nanofibers. Adv. Mater. **11**(8), 655–658 (1999)

Terrones, M., Filho, A.G.S., Rao, A.M.: Doped carbon nanotubes: synthesis, characterization and applications. In: Jorio, A., Dresselhaus, G., Dresselhaus, M.S. (eds.) Carbon Nanotubes, Topics in Applied Physics, vol. 111, pp. 531–566. Springer, Berlin (2008)

Terrones, H., Lv, R., Terrones, M., Dresselhaus, M.S.: The role of defects and doping in 2D graphene sheets and 1D nanoribbons. Rep. Prog. Phys. **75**, 062501-1–06250130 (2012)

Thess, A., et al.: Crystalline ropes of metallic carbon nanotubes. Science **273**, 483–487 (1996)

Tian, Y.: Optical properties of single-walled carbon nanotubes and nanobuds. Doctoral Dissertations. Aalto University publication, Aalto University, Finland (2012)

Tian, Y., et al.: Analysis of the size distribution of single-walled carbon nanotubes using optical absorption spectroscopy. J. Phys. Chem. Lett. **1**(7), 1143–1148 (2010)

Tirupathi, R.C., Ashok, D.B.: Introduction to finite elements in engineering, 2nd edn. Prentice-Hall, New Jersey (1997)

To, C.W.S.: Bending and shear moduli of single–walled carbon nanotubes. Finite Elem. Anal. Des. **42**(5), 404–413 (2006)

Tombler, T.W., et al.: Reversible electromechanical characteristics of carbon nanotubes under local-probe manipulation. Nature **405**, 769–772 (2000)

Tserpes, K.I., Papanikos, P.: Finite element modeling of single-walled carbon nanotubes. Compos. Part B-Eng. **36**(5), 468–477 (2005)

Villalpando-Páez, F., et al.: Synthesis and characterization of long strands of nitrogen-doped single-walled carbon nanotubes. Chem. Phys. Lett. **424**, 345–352 (2006)

Vouris, P.A.: Molecular electronics with carbon nanotubes. Acc. Chem. Res. **35**, 1026–1034 (2002)

Wal, V., Randall, L., Ticich, T.M.: Flame and furnace synthesis of single-walled and multi-walled carbon nanotubes and nanofibers. J. Phys. Chem. B **105**, 10249–10256 (2001)

Wang, S.-D., et al.: Synthesis of carbon nanotubes by arc discharge in sodium chloride solution. Carbon **43**, 1778–1814 (2005)

Wang, C.M., Zhang, Y.Y., Xiang, Y., Reddy, J.N.: Recent studies on buckling of carbon nanotubes. Appl. Mech. Rev. **63**, 030804-1-18 (2010)

Wei, C.: Thermal expansion and diffusion coefficients of carbon nanotube-polymer composites. Nano Lett. **2**(6), 647–650 (2002)

Wei, B.Q., Vajtai, R., Ajayan, P.M.: Reliability and current carrying capacity of carbon nanotubes. Appl. Phys. Lett. **79**, 1172–1174 (2001)

Wei, D.C., et al.: Synthesis of N-doped graphene by chemical vapor deposition and its electrical properties. Nano Lett. **9**(5), 1752 (2009)

Wei, L., Yanhui, F., Jia, P., Xinxin, Z.: Effects of stone-wales defects on the thermal conductivity of carbon nanotubes. J. Heat Trans. **134**, 092401-1-5 (2012)

WenXing, B., ChangChun, Z., WanZhao, C.: Simulation of Young's modulus of single-walled carbon nanotubes by molecular dynamics. Phys. B **352**, 156–163 (2004)

Wilson, M., et al.: Nanotechnology: Basic Science and Emerging Technologies. Chapman and Hall/CRC, Boca Ratón (2002)

Wu, Y., Zhang, X., Leung, A.Y.T., Zhong, W.: An energy–equivalent model on studying the mechanical properties of single–walled carbon nanotubes. Thin wall struct. **44**, 667–676 (2006)

Xiao, J.R., Staniszewski, J., Jr., J.W.G.: Tensile behaviors of graphene sheets and carbon nanotubes with multiple Stone–Wales defects. Mater. Sci. Eng. A **527**, 715–723 (2010)

Yamashita, A.M.A.S.: Carbon nanotube-based photonic devices: applications in nonlinear optics. In: Marulanda, J.M. (ed.) Carbon Nanotubes Applications on Electron Devices, pp. 367–386. InTech, Croatia (2011)

Yao, Z., Postma, H.W.C., Balents, L., Dekker, C.: Carbon nanotube intramolecular junctions. Nature **402**(6759), 273–276 (1999)

Yengejeh, S.I., AkbarZadeh, M., Öchsner, A.: On the buckling behavior of connected carbon nanotubes. Appl. Phys. A. 1–10. http://link.springer.com/article/10.1007/s00339-013-7999-2 (2014)

Yengejeh, S.I., AkbarZadeh, M., Öchsner, A.: Numerical charaterization of the shear behavior of hetero-junction carbon nanotubes. J. Nano Res-SW **26**, 143–151 (2013)

Yengejeh, S.I., Delgado, J.M.P.Q., Lima, A.G.B.d. & Öchsner, A., 2014. Numerical Simulation of the Vibration Behavior of Curved Carbon Nanotubes. *Adv. Mater. Sci. Eng.*, 2014, pp.815340-1-9

Yu, W., Xi, W.X., Xianggui, N.: Atomistic simulation of the torsion deformation of carbon nanotubes. Model. Simul. Mater. Sci. Eng. **12**, 1099–1107 (2004)

Zhou, Z.-J., et al.: Theoretical investigation on nonlinear optical properties of carbon nanotubes with Stone-Wales defect rings. J. Mater. Chem. C **2**, 306–311 (2014)

Ziaee, S.: Torsional buckling of single-walled carbon nanotubes with multi-vacancy defects. J. Comput. Theor. Nanosci. **10**(11), 2586–2590 (2013)

Chapter 3
Nanocomposites

3.1 Introduction

The idea of composites stem from the idea that by combining two or more distinct materials one may engineer a new material with properties that are a better combination of properties of the initial components. Composites, as schematically shown in Fig. 3.1, are engineered materials that consist of two or more insoluble phases combined together; a continuous phase, known as *the matrix*, as well as interdispersed components known as the *reinforcing* phase. The matrix is typically the major constituent that provides durability for the overall composite and it can be for instance, a metallic, a ceramic or a polymer material. The reinforcing inclusions are the structure's load carriers that can be in the form of fibers, particles or flakes. This phase of the composite structure provides its stiffness and strength and hence, the properties of composites depend on the properties of each of these phases, the geometry of dispersed phase, i.e. their size, distribution and orientation, and finally the amount of each phase.

Natural composites, such as wood and bones are known to everyone. However, there are also examples of historical employment of composites since about 5000 years ago, including the use of reinforcing mud walls in houses with bamboo shoots, glued laminated wood by Egyptians in 1500 B.C., and laminated metals in forging swords in A.D. 1800. In the 1930s, modern composites were used when glass fibers reinforced resins were introduced. Boats and aircraft were built out of these glass composites, commonly called fiber-glass. Since the 1970s, application of composites has widely increased due to the development of new fibers such as carbon, boron, and aramids, as well as employment of new matrices made of metals and ceramics (Kaw 2006).

Recently, carbon nanotubes composites have attracted significant attention among scientists because of their exceptional stiffness, high strength-weight ratios and toughness combined with excellent electrical and mechanical properties.

M. Rahmandoust and M.R. Ayatollahi, *Characterization of Carbon Nanotube-Based Composites Under Consideration of Defects*, Advanced Structured Materials 39,
DOI 10.1007/978-3-319-00251-4_3

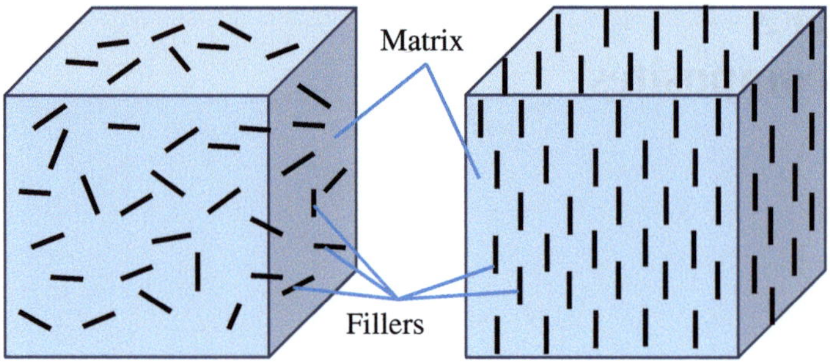

Fig. 3.1 The phases of a composite

Presentation of all these excellent properties proposes the potential of CNTs as perfect reinforcing materials in polymers. They can offer an order-of-magnitude increase in the stiffness and strength of the composite compared to typical carbon-fiber-reinforced composites.

3.2 Nanoengineered Materials

By means of advanced technologies, it has become possible to manipulate and move atoms and molecules to form new structures and thus, design new materials that are built from simple atomic-level constituents and to develop mechanical, electrical, magnetic, and other properties that are not otherwise possible. The study of the properties of these materials is termed nanotechnology; implying that the dimensions of these structural entities are in the order of a nanometer, i.e. less than 100 nm and equivalent to approximately 500 atom diameters.

In composites, therefore, if at least one of the constituent phases of the composite material, e.g. the reinforcing phase, is less than 100 nm in size, this composite is commonly termed *nanocomposite*. One example of a reinforcing material of this type is the carbon nanotube and in the future, more of our technological advances will be utilized by these nanoengineered materials.

3.3 Types of Composites

Composites are made by physically combining two or more different components to produce a combination or a blend of structural properties that are not present in any of the component individually. They can, for instance, provide greater strength and rigidity than is found in any of the separate components while being as light as possible. These engineered materials are being increasingly developed for multitude applications. For example, fiber reinforced composites are replacing metals and their alloys in many applications. Composites are designed to offer lower weight, higher stiffness and strength, lower coefficient of expansion, better resistance against fatigue, ease in manufacturing complex shapes and repairing of damaged structures, and finally better resistance to corrosion, depending on the intended function of the final device.

Based on the nature of the components, composites can be classified into three main classification, namely particle-reinforced composites; including large-particle and dispersion-strengthened composites, fiber-reinforced composites; consisting of two major types, i.e. aligned continuous/long fiber-, as well as aligned or random short fiber-composites, and finally structural composites; including laminates and sandwich panels (Callister and Rethwisch 2009).

From a technological point of view, fiber-reinforced composites are the most important composites, in which the dispersed phase is in the form of a continuous or short fiber. This type of composite can often provide high strength and/or stiffness on a weight basis, and hence the characteristics are expressed generally in terms of specific strength and specific modulus parameters, which correspond, respectively, to the ratios of tensile strength to specific gravity and modulus of elasticity to specific gravity (Callister and Rethwisch 2009). Fiber-reinforced composites are subclassified by fiber length. However, the strength of the composite depends majorly not only on the fiber length, but also to its orientation with respect to the stress direction and the efficiency of load transfer between matrix and fiber depends on the interfacial bond.

A relatively newer form of fiber-reinforced composite is the hybrid composites, which consist of two or more different kinds of fibers in a single matrix. Hybrids have better combination of properties than ordinary single fiber composites. Although a variety of fiber combinations and matrix materials are can be used in this form, carbon and glass fibers incorporated into a polymeric resin matrix makes one of the most common systems. The carbon fibers are strong and relatively stiff but expensive. Glass fibers are inexpensive and lack the stiffness of carbon. The glass-carbon hybrid is therefore stronger and tougher, but much cheaper than either of the comparable all-carbon or all-glass reinforced plastics. Principal applications for hybrid composites are lightweight land, water, and air transport structural components, sporting goods, and lightweight orthopedic components.

3.4 Structure of Composites

As mentioned earlier, the reinforcing phase and the matrix are the major constituents in all reinforced composites. Therefore, suitable selection of type, amount, dispersion density and orientation of these components plays a significant role in determining the overall characteristics of the produced composite; such as its tensile and compressive strengths, fatigue strength and failure mechanisms, electrical and thermal conductivities, specific gravity, and production costs (Kaw 2006).

3.4.1 Matrix Phase Types

The matrix phase of fibrous composites may be a metal, polymer, or ceramic. In general, metals and polymers are used as matrix materials when some ductility is desirable and in ceramic-matrix composites, the reinforcing component is added to improve fracture toughness. For fiber-reinforced composites, the matrix phase is responsible for several functions, namely binding the fibers together, acting as a medium for transmitting externally applied stresses and distributing them to the fibers. The elastic modulus of the fiber should be much higher than that of the matrix. Another important function of the matrix is protecting the fibers from surface damages resulted from mechanical abrasion or chemical reactions with the surrounding environment. Such interactions may cause surface flaws and cracks, which in turn may lead to failure at low tensile stress levels. Finally, by virtue of its relative softness and plasticity, the matrix hinders the propagation of brittle cracks from fiber to fiber. In other words, the matrix phase serves as a barrier to crack propagation and even after observing failure in some individual fibers, the composite fracture will not occur until large numbers of adjacent fibers fail and a cluster of critical size is formed.

3.4.1.1 Polymeric Matrix

Polymer is defined as a long chain of molecules, containing one or more repeating units of atoms, which have similar chemical structure but not necessarily equal length. These molecular units are joined together by strong covalent bonds to make the polymeric material. Familiar plastic and rubber materials are the commonly known polymers. In solid state, these molecules are either positioned in a random fashion in space, making amorphous polymers, or in a mixture of random and orderly folded fashions, making semi-crystalline polymers (Kaw 2006). Among different types of polymeric matrices, thermoplastic and thermoset polymer matrices make the two major categories. In thermoplastic polymers, individual molecule units are linear in structure, without any chemical linking between them, whereas thermoset polymers consist of molecules, which are chemically joined together by cross-links, forming a rigid three-dimensional network structure during polymerization reaction, which cannot be easily melted or reshaped under heat and pressure (Kaw 2006).

Composites consisting of a polymer resin as the matrix, with fibers as the reinforcement medium are known as *polymer-matrix composites* (PMCs). Due to their room-temperature properties, ease of fabrication, and cost, PMCs are used in the greatest diversity of composite applications. As the matrix often determines the maximum service temperature, depending on the targeted application, different types of polymers are employed to produce PMCs (Callister and Rethwisch 2009).

The most widely utilized and least expensive polymer resins are the polyesters and vinyl esters, used primarily with glass fibers to make composites. The epoxies are more expensive and due to their better mechanical properties and resistance to moisture, they are also utilized extensively in PMCs for aerospace applications. However, high-temperature thermoplastic resins are considered as better potentials for future aerospace applications. For high-temperature applications, polyimide resins are employed with a continuous-use, upper-temperature limit of approximately 230 °C (Callister and Rethwisch 2009).

3.4.1.2 Metal Matrix

Metals have high modulus and yield strength which candidate them for applications requiring high transverse strength and compressive strength. Compared to polymer matrices, another important advantage of a metal matrix is its long-term resistance in high temperatures, and enduring a variety of mechanical and thermal treatments, allowing them to be plastically deformed and strengthened. However, metals have some disadvantages as well, such as high specific weight, high melting points, making them hard to process. Furthermore, they have tendency toward corrosion at the matrix/fiber interface, as another weakness (Kaw 2006).

Composites with ductile metal matrix are termed as metal-matrix composites (MMCs). MMCs may be utilized at higher service temperatures than their base metal counterparts. In addition, with the help of the reinforcement, improved specific stiffness, specific strength, abrasion resistance, creep resistance, thermal conductivity, and dimensional stability would be achieved in MMCs. Some of the advantages of MMCs over the PMCs include their higher operating temperatures, nonflammability, and greater resistance to degradation by organic fluids. However, metal-matrix composites are much more expensive than PMCs, making their utilization somewhat more restricted (Callister and Rethwisch 2009).

Variety of metals are employed as matrix material; the superalloys, and alloys of aluminum, magnesium, titanium, and copper with particulates, as well as both continuous and discontinuous fibers, and whiskers reinforcements may be used. Carbon, silicon carbide, boron, aluminum oxide, and the refractory metals are used in MMCs as the continuous fiber materials. On the other hand, silicon carbide whiskers, chopped fibers of aluminum oxide and carbon, and particulates of silicon carbide and aluminum oxide are examples of common discontinuous reinforcements used in MMCs.

Automobile manufacturers use MMCs in some engine components, consisting of an aluminum-alloy matrix that is reinforced with aluminum oxide and carbon fibers. This MMC is light in weight and resists wear and thermal distortion. In addition, MMCs are utilized in driveshaft, extruded stabilizer bars, and forged suspension and transmission components as well. The aerospace industry also uses advanced aluminum alloy metal-matrix composites with boron fibers as the reinforcement for the Space Shuttle Orbiter, and with continuous graphite fibers for the Hubble Telescope. The refractory metals such as tungsten are employed as fiber reinforcement to enhance the high-temperature creep and rupture properties of some of the superalloys, e.g. Ni- and Co-based alloys. These composites have excellent high-temperature oxidation resistance and impact strength to provide higher operating temperatures and better efficiencies for turbine engines as well (Callister and Rethwisch 2009).

3.4.1.3 Ceramic Matrix

Although ceramic materials are inherently resilient to oxidation and deterioration at elevated temperatures, unlike most metals, their fracture toughness values are low and typically lie between 1 and 5 MPa. Values reported for facture toughness of most metals normally start from 15 MPa to greater than 150 MPa. However, by development of a new generation of ceramic-matrix composites (CMCs), fracture toughness of ceramics have been improved significantly to values between about 6 and 20 MPa. This improvement results from interactions between advancing cracks and dispersed reinforcement phase particles. Crack initiates normally within the matrix phase, but it is hindered by the particles, fibers, or whiskers and this way, the fracture toughness of CMCs is improved (Callister and Rethwisch 2009).

As a promising toughening technique, a phase transformation is employed to arrest the propagation of cracks by using small particles of partially stabilized zirconia as dispersed reinforcement phase within the ceramic matrix. The common particles include Al_2O_3 or ZrO_2, whereas CaO, MgO, Y_2O_3, and CeO are used as stabilizers. The stress field in front of a propagating crack causes these particles to undergo transformation to the stable monoclinic phase and a slight particle volume increase. Therefore, compressive stresses are established on the crack surfaces near the crack tip that tend to pinch the crack shut, and hence, arresting its growth.

Another developed toughening technique involves utilization of ceramic whiskers, often SiC or Si_3N_4. These whiskers may hinder the crack propagation by deflecting crack tips, forming bridges across crack faces, absorbing energy during pull-out as the whiskers debond from the matrix, and/or by causing a redistribution of stresses in regions adjacent to the crack tips. In general, increasing fiber content improves strength and fracture toughness. Furthermore, there is a considerable reduction in the scatter of fracture strengths for whisker-reinforced ceramics relative to their unreinforced counterparts. These CMCs exhibit improved high-temperature creep behavior and resistance to thermal shock (Callister and Rethwisch 2009).

Ceramic matrix composites may be fabricated using hot pressing, hot isostatic pressing, and liquid phase sintering techniques. Relative to applications, SiC whisker-reinforced alumina composites are being utilized as cutting tool inserts for machining hard metal alloys; tool lives for these materials are greater than for cemented carbides. Some of ceramic-matrix composites would be ideal candidates for use in high-temperature and severe-stress applications, specifically for components in automobile and aircraft gas turbine engines (Callister and Rethwisch 2009).

3.4.2 Reinforcing Phase Types

As an important characteristic of most materials, especially brittle ones, it is known that by reducing the diameter, the material becomes much stronger than the bulk material. Furthermore, the probability of the presence of a critical surface flaw that can lead to fracture diminishes by decreasing the volume of the material. These features are considered advantageous in the science and technology of fiber-reinforced composites.

According to the diameter, fibers are categorized into three different groups, namely whiskers, fibers, and wires. *Whiskers* are very thin, highly-perfect single crystals that have extremely large length-to-diameter ratios, and exceptionally high strength. However, due to their high cost, the strong whiskers are not utilized extensively as a reinforcement medium. Moreover, it is difficult and often impractical to incorporate whiskers into a matrix. Whisker materials include graphite, silicon carbide, silicon nitride, and aluminum oxide. The second group, i.e. *fibers*, is either polycrystalline or amorphous polymers or ceramics. Polymer aramids, glass, carbon, boron, aluminum oxide, and silicon carbide are categorized in this group. Finally, *wires* have relatively larger diameters, compared to whiskers and fibers. Typical wires include steel, molybdenum, and tungsten.

The materials generally used for reinforcing fibers have high tensile strengths and the adhesive bonding forces between fiber and matrix plays a very important role in achieving minimized fiber pull-out and maximizes stress transmittance from the weak matrix to the strong fibers. In fact, bonding strength is one of the most important considerations in the choice of the matrix-fiber combination and the ultimate strength of the composite depends very much on the strength of this bond (Callister and Rethwisch 2009). For instance, in PMCs carbon, glass and aramid fibers are the most common reinforcing phases employed. Various types of fibers employed in the industry of manufacturing composites are explained further in this section.

3.4.2.1 Carbon Fibers

The terms graphene and carbon are often interchangeably used in the composite community. Graphene is the basic structure of graphite and carbon nanotubes and the isolation of this narrow sheet of carbon atoms, which is only one atom thick by Professors Andre Geim and Konstantin Novoselov was so important that the two scientists were awarded the Nobel Prize in Physics in 2010 (Nobel-Prize 2010). Carbon fibers are used as filler in a continuous matrix of a polymer, which is often an epoxy resin. The polymer matrix is responsible of transferring stresses between the fibers and protecting their surface from mechanical abrasion. In addition, the matrix provides lateral support against the possibility of fiber buckling under compression (Kaw 2006).

Carbon fiber reinforced polymer (CFRP) composites are relatively expensive. However, when lightness and high strength are of dominant importance, they are cost effective for the following their many good advantages.

1. They have extraordinarily high specific strength, i.e. tensile strength per weight, and specific tensile modulus at elevated temperatures.
2. Their coefficient of thermal expansion is very low and they have high fatigue strength.
3. At room temperature, carbon fibers are not affected by moisture or a wide variety of solvents, acids, and bases.
4. These fibers exhibit a diversity of physical and mechanical characteristics, allowing composites incorporating these fibers to have specific engineered properties.
5. Recent manufacturing processes of CFPRs are relatively inexpensive and cost effective.
6. They behave anisotropic and have a high longitudinal stiffness, due to alignment of the basal planes parallel to the fiber axis (Callister and Rethwisch 2009).

Carbon fibers are sometimes classified based on their tensile modulus to four classes of standard, intermediate, high, and ultrahigh moduli carbon fibers. Fibers are in some applications coated with a protective epoxy to improve their adhesion with the polymer matrix.

Carbon-reinforced polymer composites are currently being utilized extensively in sports and recreational equipment, e.g. fishing rods and golf clubs, filament-wound rocket motor cases, pressure vessels (Callister and Rethwisch 2009). Furthermore, these reinforcing materials are mostly very appropriate options in both military and commercial aerospace industry, where weight-savings is considered to be more critical than lowering costs (Kaw 2006). The fuselages of the newest commercial airplanes, i.e. Boeing 787 and Airbus A350, are constructed mostly from CFRP composites. In addition, aircraft brakes are required to absorb considerable quantities of energy rapidly without mechanical failure. The usual construction is based on multiple rotating and stationary discs, which can reach surface temperatures of up to 3000 K and hence the disc material must be of an excellent thermal and shock resistance and high-temperature strength; it should also have a good thermal conductivity. CFRPs are ideal material for this purpose (University-of-York 2013).

In addition, CFRPs have also been used for many years to construct the body of F1 racing cars, giving drivers greater protection even in crashes at over 300 km/h. They are now being used in part of some luxury cars like in the Mercedes Benz range, in the roof of the GM Corvette ZR1, and as protective gear for motor cyclists. They are also employed in large structures such as in bridges and as a strengthening wrap around the parts that require more strength (University-of-York 2013).

3.4.2.2 Glass Fibers

Glass fibers are majorly made from silica (sand), sodium carbonate and calcium carbonate. The materials are heated to about 1700 K in a furnace and then extruded directly from the furnace to produce filaments (University-of-York 2013). Glass fiber-reinforced polymer (GFRP) composites are produced in the largest quantities and the reason behind their popularity lies behind the following major advantages:

1. They can be easily drawn into high-strength fibers from the molten state.
2. Using a wide variety of composite-manufacturing techniques, GFRPs can be produced with very low costs.
3. Glass fibers are usually isotropic and relatively strong, and when embedded in a polymer matrix, it produces a composite with very high specific strength.
4. Chemical resistance and excellent insulating properties renders the composite useful in a variety of corrosive environments (Callister and Rethwisch 2009).

However, they have some disadvantages as well, such as low tensile modulus, relatively high specific gravity, low fatigue resistance, high hardness and sensitivity to abrasion that decrease their tensile strength (Kaw 2006). In addition, application of most fiberglass composites is limited to temperatures below 200 °C and at higher temperatures, they begin to flow or to deteriorate. The service temperature of GFRPs, however, may be extended to 100 °C higher by using high-purity silica for the fibers and high-temperature-durable polymers such as the polyimide resins (Callister and Rethwisch 2009).

Surface characteristics of glass fibers are extremely important because even minor surface flaws can affect the tensile properties of the fiber deleteriously by showing weakened bonding with the matrix. Hence, newly drawn fibers are normally coated during drawing with a "size", i.e. a thin layer of a substance that protects the fiber surface from damage and undesirable environmental interactions. This size is ordinarily replaced with a "coupling agent" prior to composite fabrication to produce a better chemical bond with the matrix (Callister and Rethwisch 2009).

Many fiberglass applications are familiar, like automotive and marine bodies, plastic pipes, storage containers, and industrial floorings. Based on their application, glass fibers are categorized into two important groups of so-called *E-glass* fibers, with boosted electrical properties, and *S-glass* fibers, named so due to their high tensile strength (Kaw 2006).

GFRPs are used widely in manufacturing of boats. While large ships are usually constructed in steel, over 80 % of marine hulls, that are less than about 40 m in length, are made of glass fiber reinforced polymer. This is due to their cheaper manufacturing process and easier to maintenance. Additionally, there are certain applications in which the magnetic, electrical or thermal properties of GFRPs are advantageous, e.g. in non-magnetic minesweepers to avoid activation of mines. Glass-phenolic resins are used to improve protection for armored vehicles as well (University-of-York 2013).

3.4.2.3 Aramid Fibers

Introduced in the early 1970s, aramids are chemically polymeric aromatic amide fibers. These high-strength, high-modulus materials with outstanding specific strength or strength-to-weight ratio, which is superior to metals, are the most widely used anisotropic organic fibers. However, although their tensile strength and stiffness are very high in the axial direction of the fiber, the values of the mentioned properties are relatively low in the transverse direction and low compressive strength is the major disadvantage of these types of reinforcing fibers (Kaw 2006; Callister and Rethwisch 2009).

There are a number of aramid materials including Kevlar™ and Nomex™ as two of the common ones. Kevlar® is very difficult to process as its only solvent is concentrated sulfuric acid. It is made from lightweight atoms but is very strong and flexible, due to the way that its polymer chains are aligned, i.e. flat sheets of molecules held together by hydrogen bonds that can stack up along the fiber axis (University-of-York 2013). In Kelvar™, the rigid molecules are aligned in the direction of the fiber axis as liquid crystal domains during the synthesis process, making mechanically high longitudinal tensile strengths and tensile moduli materials. However, they are relatively weak in compression. In addition, this material has high toughness, impact resistance, and resistance to creep and fatigue failure.

Aramids are resistant to combustion and stable to relatively high temperatures; the temperature range over which they retain their high mechanical properties is from −200 to 200 °C. Furthermore, the aramid fibers are susceptible to degradation by strong acids and bases; however, they are relatively inert in other solvents and chemicals. These fibers are mostly used as reinforcement within polymer matrices, like epoxies and polyesters (Callister and Rethwisch 2009).

Aramid fiber reinforced composites are particularly useful where energy has to be absorbed and dissipated, and they are also able to resist abrasion. These composites have been used widely in aviation, for helicopter rotor blades, in sport goods, to make tennis, badminton and squash racquets, and in boats such as kayaks and dinghies (University-of-York 2013). Other applications of aramid composites are in ballistic products, such as bulletproof vests and armor, tires, ropes, missile cases, pressure vessels, and as a replacement for asbestos in automotive brake and clutch linings, and gaskets (Callister and Rethwisch 2009).

3.4.2.4 Other Fibers Types

Glass, carbon, and the aramids, as mentioned earlier, are the most common fiber reinforcements incorporated typically in polymer matrices. However, other fibers such as boron, silicon carbide, and aluminum oxide and other ceramics are used also employed in composite industry (Callister and Rethwisch 2009).

Boron fiber-reinforced polymer composites have been used in military aircraft components, helicopter rotor blades, and some sport goods. These fibers were one of the first high performance fibers available for use as reinforcing phase in composite materials. The diameter of boron fibers is in the range of 0.1–0.2 mm, which is an order of magnitude larger than that of glass, aramids, or graphite fibers. Due to their large diameter and high stiffness restricts their bend radius greatly, offering high resistance to buckling, which in turn contribute to an excellent compressive performance of boron-reinforced composites (Kaw 2006).

Ceramic fibers, including silicon carbide, aluminum-boron-silica and aluminum oxide fibers are utilized in tennis rackets, circuit boards, military armor, and rocket nose cones (Callister and Rethwisch 2009). These fibers are exceptionally resistance to extremely high temperatures, while still maintaining competitive structural properties. Being incorporated in metal and ceramic matrices, ceramic fibers are suitable reinforcing options for metal matrices in which boron and carbon fibers exhibit adverse reactivity (Kaw 2006).

3.4.3 Continuous-Fiber Composites

Known to have high tensile strength, composites are used in many applications for their superior mechanical properties. The mechanical behavior of this type of composites, however, depends on various factors, namely, the stress-strain response of both, fibers and matrix, phases, volume fraction of phases, as well as the direction in which the load is applied. In addition, if the fibers are aligned and long or continuous, as schematically illustrated in Fig. 3.2, their properties will be highly dependent on the direction in which they are measured and hence, strongly anisotropic (Callister and Rethwisch 2009).

Initially, fibers and matrix deform elastically together, but after the initial stage, normally the matrix yields and deforms plastically, whereas the fibers continue to stretch elastically. It is because the tensile strength of the fibers is considerably higher than the yield strength of the matrix. Further load application leads to composite failure and the fibers start to fracture. However, since not all fibers

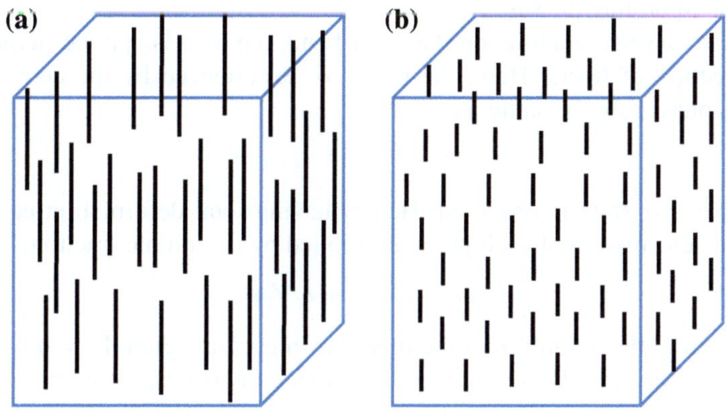

Fig. 3.2 **a** Long-fiber, and **b** short-fiber composites

fracture simultaneously, the failure is not catastrophic (Callister and Rethwisch 2009). Furthermore, although the fractured fibers are shorter than the initial fibers, they are still capable of sustaining load and hence, the matrix continues to deform plastically.

In the elastic region, when there is good bonding between long fibers and the matrix of the composite, external loaded applied in the direction of fiber alignment, deforms both matrix and fibers the same way, i.e. an *isostrain* situation. Therefore, the total load carried by the composite, F_c, is equal to the sum of the loads carried by the matrix, F_m, and the fibers, F_f.

$$F_c = F_m + F_f \tag{3.1}$$

Having $F_i = \sigma_i A_i$, for any respective stress σ, and cross-sectional area A, and dividing Eq. (3.1) by the total cross-sectional area of the composite, A_c, we will have

$$\sigma_c = \sigma_m (A_m/A_c) + \sigma_f \left(A_f/A_c\right) \tag{3.2}$$

where A_m/A_c and A_f/A_c are the area fractions of the matrix and fiber phases, respectively. If the lengths of the composite, matrix, and fiber are equal, then the fraction would give us the volume fraction of the matrix, V_m and the fibers, V_f, respectively.

$$\sigma_c = \sigma_m V_m + \sigma_f V_f \tag{3.3}$$

Looking back at the isostrain state of composite phases in the elastic region, i.e.

$$\varepsilon_c = \varepsilon_m = \varepsilon_f, \tag{3.4}$$

when dividing each term in Eq. (3.3) by its pertaining strain and by having Young's modulus $E_i = \sigma_i/\varepsilon_i$ in the elastic region, the moduli of elasticity for long aligned fiber composites in longitudinal direction would be

$$E_{c-longitudinal} = E_m V_m + E_f V_f, \tag{3.5}$$

which highlights the influencing role of volume fraction and modulus of the composite's incorporating phases.

For the transverse loading, on the other hand, the load is applied perpendicular to the direction of fibers. Here, the stress, σ, experienced by the composite and both of its phases are the same, i.e.

$$\sigma_c = \sigma_m = \sigma_f, \tag{3.6}$$

and the *isostress* state is observed. Hence, having total deformation equal to the sum of the deformation of each phase multiplied by its volume fraction,

$$\varepsilon_c = \varepsilon_m V_m + \varepsilon_f V_f, \tag{3.7}$$

and considering the definition of the modulus in elastic region, $E_i = \sigma_i/\varepsilon_i$, then the transverse elastic modulus for aligned long fiber composites, would be

$$1/E_{c-transverse} = V_m/E_m + V_f/E_f. \tag{3.8}$$

3.4.4 Short-Fiber Composites

Recently, short-fibers composites have become increasingly popular because of their easy ability of compatibility to conventional techniques of manufacturing and low-fabrication cost. Since 1963, many researchers have been studied widely on the effective macroscopic behavior of composite materials from a theoretical viewpoint (Hill 1963; Hashin 1983; Mura 1987; Aboudi 1991). Fukuda and Kawata predicted the Young's modulus of short fiber-reinforced materials under plane stress. Firstly, they analyzed the mechanism of load transfer between fiber and matrix in a single short fiber model using the theory of elasticity. Next, by applying the obtained results they estimated Young's modulus of composite materials assuming that the interactions between fibers are negligible. According to their results, Young's modulus of unidirectional reinforced materials with short fibers is smaller than that with continues fibers. In addition, they reached to the conclusion that misalignment of fibers leads to the reduction in the values of Young's modulus; it means that the Young's modulus decreases when the value of fiber orientation changes from 0° to 90° (Fukuda and Kawata 1974). Chiang studied the tensile strength of short fiber-reinforced composites by the statistical theory of the strength of fiber bundles. His research considered composites with unidirectional, misaligned and random-fiber directions. Finally, after the prediction of the tensile strength it was concluded that if the strength of the matrix was ignored, then a composite with unidirectional fibers would be twice as strong as a composites with planer random fibers (Chiang 1994).

Chen and Cheng investigated the effective moduli of composites containing disoriented fibers based on the Eshelby-Mori-Tanaka theory. These moduli were given in terms of the length to diameter aspect ratio, the volume fraction, and the orientation distribution parameter of the short fibers. According to their study, the fiber orientation distribution and aspect ratio had more significant effects on composite longitudinal Young's modulus than the fiber volume fraction did within the range examined. They concluded that the elastic moduli of a short-fiber-reinforced composite depended on the corresponding properties of the fiber and the matrix, i.e. volume fraction, aspect ratio, and orientation factor of the fibers (Chen and Cheng 1996). Elastic moduli will increase by increasing either the fiber volume fraction or with aspect ratio. The increase however, is much more significant for the longitudinal tensile modulus, E_{11}, in case of aspect ratio. Therefore, it is imperative to optimize the conditions for compounding and processing of composites reinforced with high aspect ratio fibers preferably.

With the rapid development of computing power, numerical simulation can be anticipated to play an increasingly significant role in characterizing short-fiber reinforced composites. Recently, Haitao and Zhenhan analyzed the mechanical properties of short-fiber reinforced composites using a new fast multipole boundary element method (BEM). The model they used in their calculation is a 3-D solid model with 100 randomly distributed and oriented short fibers that have been calculated using the fast multipole BEM. They evaluated the effects of the fiber aspect ratio and the fiber orientation on both the effective elastic properties and the interfacial stress distributions of 3-D fiber-reinforced composites using the fast multipole BEM. They considered three fiber aspect ratios, i.e. length/diameter = 3, 6, and 11, and four fiber orientation, i.e. 0°, 0°–30°, 0°–60°, and 0°–90°. According to their results, the effective young modulus E_x decreases as the fiber orientation distribution changes from the unidirectional mode (0°) to the random mode (90°). For a fixed fiber orientation distribution, E_x increases with increasing fiber's aspect ratio. Furthermore, the differences between the E_x values for the different aspect ratios tend to decrease with increasing fiber orientation spreads, and almost vanish for the maximum fiber orientation spread of 90°. Finally, they investigated that longer fibers result in higher stress concentrations than shorter ones (Haitao and Zhenhan 2007).

Among all the variety of different short-fibers that can be used as a nanocomposite's reinforcing phase, CNTs have shown to be promising candidates for their very specific mechanical and physical properties. Carbon nanotube-based nanocomposites, i.e. composite materials in which carbon nanotubes are used as the composite's reinforcing phase, are therefore very much interesting for scientists and scholars, for the many outstanding and remarkable applications that they can contribute to the world of science and industry. They are expected to influence many fields in terms of technology and industry. They will have applications in many diverse fields such as energy, signal processing, medicine, biotechnology, information technology, aerospace, agriculture, and environment (Wang et al. 2010). These outstanding materials can be used as stand-alone nanomaterials or as reinforcements in composites for a wide variety of application. Therefore, several detailed studies have been conducted to explore different properties of carbon nanotubes and carbon nanotube based composite materials.

Although aligned fiber composites demonstrate five times stronger properties that is anisotropic in nature, compared to random orientation of discontinuous or short fibers, depending on the direction of loads, there are applications involving multidirectional stresses that would require short fibers (Callister and Rethwisch 2009). As it will be explained later in this chapter, rule of mixture is a common technique for predating the behavior of this type of composite; giving elastic modulus equal to

$$E_{\text{c-short}} = KE_f V_f + E_m V_m. \tag{3.9}$$

where K, the efficiency parameter depends on V_f and the ratio of E_f/E_m. Schematic illustration of (a) aligned- and (b) randomly-oriented short finer composites is provided in Fig. 3.3.

In the transverse direction, fibers are practically incapable of reinforcing the composite, and fracture happens usually at rather low tensile stresses. The reinforcement efficiency factor is a value, introduced in 1964, which varies between one for stresses applied parallel to the fibers and zero for transverse load, in a way that any arrangement of short fibers lays between these two extremes (Krenchel 1964). For instance, reinforcement efficiency for fibers randomly and uniformly distributed within a specific plane and load applies in any direction in the plane is

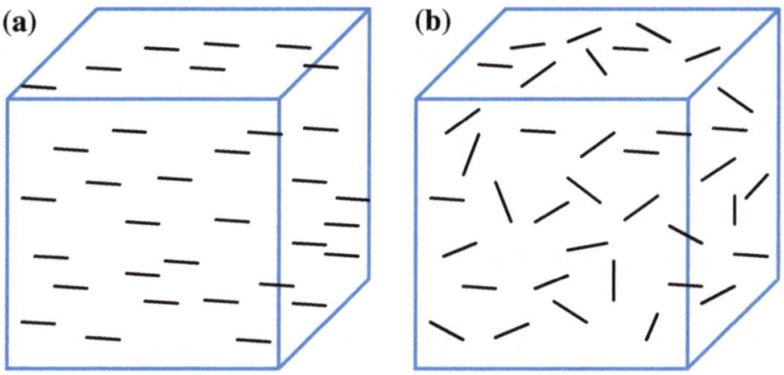

Fig. 3.3 **a** Parallel and **b** Random orientation of CNTs inside matrix

equal to 3/8 and in case of fibers randomly and uniformly distributed within three dimensions in space and stress applies in any direction is 1/5.

In comparison, since the production rates of short-fiber composites are faster and various complex composite shapes can be fabricated with fibers that cannot be achieved with continuous-fiber composites, short-fiber reinforcement is a more favorable choice in industry. In addition, costs of manufacturing this type of composites are much lower than that of continuous fiber composites.

3.5 Carbon Nanotube-Based Composites

Composite materials mostly show more significant advantageous properties compared to monolithic materials. Monolithic metals and their alloys cannot always meet the demands of today's advanced technologies and performance requirements. It means that by incorporating reinforcements into a, for instance, metallic, ceramic or polymer matrix, the properties of the matrix improves toward higher mechanical strength, more significant temperature stability and better chemical durability. The existence of reinforcing elements can boost the structure's physical and chemical properties significantly.

As mentioned earlier, the discovery of CNTs offered exciting opportunities for emerging applications in most areas of science and engineering. CNTs can be considered as carbon fibers that have break strengths of as high as 200 GPa and elastic moduli of about 1 TPa. In addition, these nano materials have approximately 500 times more surface area per gram compared to typical carbon fibers and aspect ratio of around 103, making them suitable options for employment as reinforcing phase for polymer matrices (Khare and Bose 2005).

Hence, many studies have focused on characterizing the properties CNT-based composites experimentally or by simulating their molecular structure using different computational method. Carbon nanotubes, as reinforcing materials for polymer-, metal- or ceramic-matrix composites, are expected to help obtaining light-weight structural materials with enhanced mechanical, electrical and thermal properties (Sie 2009). Among these investigations, special emphasis will be given here to those that considered the effect of *imperfections* in the structure of CNTs and CNT-based composites.

3.5.1 Mechanical Properties

A great deal of theoretical work has been carried out to predict mechanical properties of fiber reinforced composites and some of those models are quite sophisticated for nanocomposites (Tucker and Liang 1999). The properties of composites are strongly affected by a number of factors, which are as follows:

1. Geometry and properties of reinforcement elements;
2. Concentration and volume fraction of reinforcing material with respect to the matrix;
3. Properties of continuous phase, i.e. the matrix, in the composite;
4. Homogenous dispersion of reinforcement in the polymer matrix; and
5. Interfacial strength between reinforcement and polymer matrix.

By controlling the concentration of reinforcements, the properties of the new material can be tailored to suit the desired mechanical requirements. By varying the type and distribution of the reinforcing phases in the composites, it is possible to obtain a wide range of mechanical and physical properties and hence to optimize the structure and its performance and expect different interaction with the surrounding in order to be used for different devices and expect different interaction with the surrounding.

Remarkable mechanical properties of carbon nanotubes are an important part of their engineering value. Young's modulus of single- and multi-walled carbon nanotubes are about 1 TPa and 0.3–1 TPa respectively, whereas their strength is reported as to be 50–500 GPa and 10–60 GPa, respectively. Furthermore, CNTs have many other interesting properties such as ordered structure with high aspect ratio, ultra-light weight, and high mechanical strength. As a result of these exceptional properties, both theoretical and experimental results introduces these advanced structured materials as good candidates to be utilized as reinforcing elements in nanocomposites (Ajayan 1999; Goze et al. 1999; Dresselhaus et al. 2001; Xie et al. 2005). The combination of these characteristics makes CNTs and CNT-based nanocomposites unique materials with the potential for many emerging nanoengineering applications, including biomedical engineering but the superior mechanical properties of CNTs alone do not guarantee the nanocomposite's superior strength, stiffness, and fracture toughness nonetheless (Xie et al. 2000; Iijima 2002; Bianco et al. 2005). Furthermore, the empty space inside carbon nanotubes can be used to incorporate atoms and molecules in order to generate novel compounds, nanostructured materials, and nanocomposites. The possibilities of achieving nearly one-dimensional nanostructures by filling the hollow

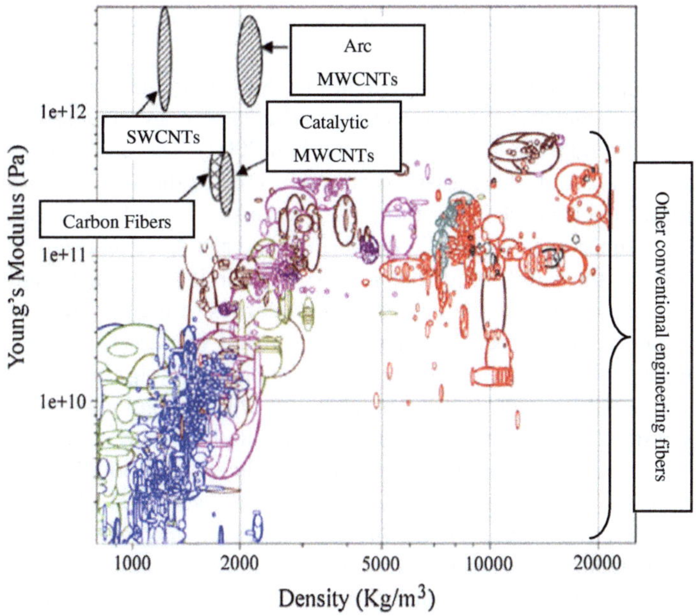

Fig. 3.4 Young's modulus versus density for different engineering materials (modified from Esawi and Farag 2007)

nanotubes have generated a great deal of interest in recent years as well (Ajayan et al. 1993; Ajayan and Iijima 1993).

There have been numerous studies performed involving the addition of CNTs into various matrices, with the purpose of improving the material properties, due to their superior strength and stiffness. Ajayan team presented one of the first studies on polymer/CNT composites (Ajayan et al. 1994) and since then, many computational, theoretical and experimental studies have been devoted to investigate the elastic properties of CNT-based composites. The tensile strength of an individual carbon nanotube is over 150 GPa, which means that they are a hundred times stronger than steel, though three to five times lighter (Esawi and Farag 2007). Figure 3.4 shows a comparison between the tensile modulus versus density for various engineering materials including the conventional engineering materials and carbon nanotubes.

3.5.1.1 Interfacial Adhesion

Enhancement of the modulus of the composites reflects the transfer of stress from the polymer matrix to the high-modulus embedded CNTs. Thus, *strong* interfacial adhesion between the polymer and CNT is preferred for stiffening nanocomposites. Enhancement of the strength of composites reflects strong interaction between the CNT and polymer, but strength is also profoundly affectedly the presence of defects such as voids or agglomerate inclusions, which serve to initiate failure by stress concentration (Bubert et al. 2003).

Uniform dispersion and well alignment of CNTs within the polymer matrix, as well as improved matrix/nanotube wetting and adhesion are critical factors in the processing of the nanocomposites (Lau and Hui 2002). Therefore, processing can have a major impact on the measured strength. Due to the very high aspect ratio of nanotubes, which causes extremely high viscosity in melt or dissolved polymers, the nanotubes usually exist in the form of micron-sized agglomerates, which adversely affects their uniform dispersion. As a result, it limits their loading in polymers and hence influences the mechanical properties of the resulting composites. If they are poorly dispersed, the nanotubes will fail by separation of the bundle rather than by failure of the nanotube itself, leading to a significant reduction of strength. It is worth considering that breaking up of these agglomerates by conventional processing technique seems to be rather difficult (Bubert et al. 2003).

Numerous researchers have reported that the interfacial bonding in CNT/polymer composites is lower-than-predicted (Weisenberger et al. 2003; Nan et al. 2003). From micromechanics point of view, however, the interface between the CNT and surrounding polymer matrix is essentially an exposed graphene sheet with a weak inter-planar interaction that provides its solid lubricant quality and resistance to matrix adhesion. This fact is exaggerated by the chemically inert nature of graphene structures (Khare and Bose 2005). Therefore, considering that stress is transferred through shear stress build-up at this interface from the matrix to the CNT, the interface between nanotubes and the polymer matrix must be chemically or physically modified in order to promote the stress transfer between matrix and tubes. When the interfacial adhesion between the phases is weak, the nanotubes behave as holes, inducing local stress concentrations, and the benefits of the CNT properties are lost (Lau and Hui 2002). Otherwise, the strong adhesion leads to excellent transfer of stress in the structure and better mechanical strength, consequently. The interaction between tube components of nanocomposites at the nanometer scale enables them to act as molecular bridges in the polymer matrix. This is the basis for enhanced mechanical properties of the nanocomposite as compared to conventional microcomposites. If the above said problems are sorted out, enhanced properties of nanocomposites with full utilization of the properties of CNTs will be achieved.

3.5.1.2 Computational Studies

Many techniques have been employed to explore the material properties of CNT-based composites. As the experimental characterization of nanocomposites is expensive and applying analytical solution to characterize these structures at nanoscale is a difficult task, emerging computational methods have been increasingly applied in characterization of nanocomposites. Molecular dynamics, continuum mechanics and combinations of these two main methods have been employed for the characterization purpose.

As a result of the high computational power that MD simulations requires, this type of computational method imposes restrictions, specifically in terms of the studied model's dimensions and timescale. For instance, the majority of available MD simulations of CNT composites are limited to models with a single CNT inside a supporting matrix; whereas, in a real composite, CNT fibers can have different shapes and sizes or they can be straight or curved, short or long, aligned or arbitrary oriented, and randomly distributed. All these factors make it very difficult to use the MD approach in investigations of the mechanical properties of CNT composites. Continuum mechanics approaches, on the other hand, can fill this gap and make the task easier by introducing a good computational efficiency in characterizing large-scale complex models.

A continuum mechanics approach was first reported to be used for characterizing the stress distribution and effective elastic properties of CNT composites (Pipes and Hubert 2002). They applied the anisotropic elasticity theory to study the effective mechanical properties of CNT-based composites. Liu and Chen obtained the effective mechanical properties of CNT based composites using the boundary element and the finite element method. They used a squared and a cylindrical *representative volume element* (RVE), with one or multiple, short or long single-walled carbon nanotubes. In order to evaluate the effective elastic properties of the composite, the BEM was used (Liu and Chen 2003a), Liu and Chen applied the advanced multi-domain to the squared RVE model and studied the overall deformation, effective Young's modulus and load transfer mechanism of the RVE. Later in their other research, a cylindrical RVE with three loading cases for a single-walled carbon nanotube in a matrix material was studied to obtain its effective Young's modulus and Poisson's ratio using the FEM (Liu and Chen 2003b). They extended their previous study on squared RVE for evaluation of the effective materials properties of CNT-based composites, based on a FEM approach (Chen and Liu 2004). Finally, based on comparison, it was concluded that the squared RVEs give more accurate results compared to the cylindrical one.

Fereidoon considered a three-dimensional model of a CNT in a cubic representative volume element first and then, the model was reproduced by replacing the embedded 3 D CNT with an equivalent one dimensional beam element, using the finite element modeling. They studied the effect of changing fiber volume fraction on the composite's Young's modulus. Results show that by increasing fiber volume fraction, the elastic modulus will increase and the trend is approximately linear (Fereidoon et al. 2008). Shokrieh and Rafiee studied the effect of nanotube length on the reinforcement efficiency of carbon nanotubes in composites. They concluded that mechanical properties of nanotubes, interfacial bonding and orientation of CNTs in a matrix are the most influencing factors in the performance of carbon nanotube-reinforced composite. The orientation of CNTs in a surrounding polymer affects the performance of the composite both at macro- and nano- scales. The load transfer mechanism from matrix to CNTs is the main governing factor that characterizes the efficiency of reinforcement. The lengths of CNTs also have a significant effect on the reinforcement efficiency (Shokrieh and Rafiee 2010).

The *effective reinforcing modulus* (ERM) of a wavy embedded nanotube was determined both by Farsadi et al. and by Fisher et al. (Farsadi et al. 2013; Fisher et al. 2003). A combination of the finite element and micromechanical methods was used by Fisher research team. They used ERM within a multiphase micromechanical model to estimate the effective modulus of a polymer reinforced by a distribution of *wavy* nanotubes. The results show that even a very small curvature in the structure of the embedded CNTs reduced the effective reinforcement ability of them remarkably, compared to straight nanotubes (Fisher et al. 2003). Farsadi, on the other hand, used a three-dimensional finite element model to investigate the effect of volume fraction and waviness on Young's modulus and Poisson's ratio of composites reinforced with waved carbon nanotubes. According his results, nanotube waviness causes a decrease in the longitudinal and transverse Young's modulus of composites, compared to the straight nanotube reinforcement but the change in the value of transverse Young's modulus, i.e. E_{yy}, is less remarkable than the longitudinal Young's modulus, E_{xx}. Furthermore, the fiber curvature was proved not to have much effect on Poisson's ratio. The fiber volume fraction was also investigated on the longitudinal Young's modulus and major Poisson's ratio (Farsadi et al. 2013).

Odegard developed a new method as an equivalent continuum modeling technique for improving constitutive models for single-walled carbon nanotube reinforced composites. Using molecular dynamics simulations, atomic level interactions at nanoscale, interfacial characteristics of the nanotube and the surrounding polymer matrix were considered in this method. After that, they developed their equivalent continuum model to obtain the mechanical properties that reflect the local polymer/nanotube interactions. This model has been used to determine the constitutive properties of single-walled carbon nanotube/polymer composite in aligned and randomly orientated distribution regimes, with various nanotube lengths and volume fractions (Odegard et al. 2001). Later, Liu research team used the fast multipole boundary element method, developing a new continuum model of CNT composites, in order to characterize CNT-reinforced composites. In their approach, due to their high stiffness, CNTs were treated as rigid fibers in an elastic matrix. They used two different distributions of the fibers in the RVE, uniform distribution of aligned fibers and random distribution of aligned CNT fibers. The results for the aligned random and uniform cases were reported to be very close, possibly because the extent of randomness was not remarkable enough, with respect to the uniform distribution. In all the above studies, researchers used different volume fractions of CNT fibers to evaluate the effective elastic modulus of composites, by modeling and simulating the unit cell, holding a single CNT fiber, or an RVE structure that is composed of multiple aligned unidirectional CNT fibers (Liu et al. 2005).

As mentioned before, the possibilities of achieving nearly one-dimensional nanostructures by filling hollow nanotubes have also attracted a great deal of attention. By incorporating atoms and molecules inside hollow nanotubes, brand new compounds, advanced materials and nanocomposites are produced (Ajayan et al. 1993; Ajayan and Iijima 1993; Liang et al. 2004; Sinnott et al. 2004). Filled carbon nanotubes can therefore act as reinforcements in composites. The first experimental experience of filling of single-walled carbon nanotube was reported by Ajayan and Iijima (1993). Later, Kalamkarov team, in their studies on mechanical properties of filled single-walled carbon nanotubes, considered a two-phase cylinder model. Both materials, i.e. the inside matrix and the outside SWCNT were assumed to be of two different orthotropic materials. Based on these assumptions, a finite element model was generated and the elastic modulus of the structure was calculated numerically to be equal to 0.165 TPa, where the elastic modulus of the matrix was 3.45 GPa (Kalamkarov et al. 2007).

Recently, continuum mechanics approach was employed to numerically analyze the effect of CNT reinforcement on a two phase polymer composite and compare the results with experimental investigations. The studied volume fraction of CNT varied from 0.5 to 5 % and both finite element analysis and rule of mixture theory were employed for the study. According to the obtained results, low volume fraction of CNT leaded to a large increase in elastic modulus of the composite. On the other hand, comparing the results with experimental results clearly indicates that although the large volume fraction of CNT increases the viscosity and makes the processing issues very difficult, three phase polymer composite with low volume fraction of CNT can be definitely employed in aerospace and other structure applications (Gupta and Harsha 2014).

3.5.1.3 Experimental Studies

The mechanical properties of carbon nanotube-based composite materials, mostly embedded inside polymer matrices, have been also studied by different scholars. For instance, Zhong studied a fabricated SWCNT-reinforced nanocrystalline-aluminum (SWCNT/nano-Al) composite experimentally. They found the highest hardness of the SWCNT/nano-Al composites to be equal to 2.89 GPa, after an optimum consolidation procedure, which was approximately 20 times higher than that of a coarse-grained Al, i.e. 0.15 GPa, and almost two times higher than nano-Al, i.e. 1.62 GPa, consolidated at the same temperature. The novel mechanical properties of the SWCNT/nano-Al composites indicate that SWCNTs can be used as an excellent reinforcing phase in a nano-Al matrix (Zhong et al. 2002). In another early investigation on the influence of adding MWCNT to polymer matrices, Allaoui team found that compared to the pure resin matrix samples, the Young's modulus and the yield strength of the nanocomposite increased to two- and four-times more respectively by adding 1 and 4 wt% nanotubes (Allaoui et al. 2002).

Tang studied mechanical properties of CNT/HDPE nanocomposites by a small punch test, which was adopted by Kurtz and Edidin (Kurtz et al. 1997; Edidin and Kurtz 2001). Since the loading configuration involves lateral bending and large biaxial deformation, the experimental results do not directly yield the usual mechanical properties. Focusing on the small deformation region, experimental results of the elastic modulus were obtained by comparing the initial slope of the force-displacement curve and they found a stiffness increase (Tang et al. 2003).

Incorporation of CNTs into polymer matrices also leads to significant toughening of the structure. A loading of 1 wt% MWCNT, randomly distributed inside an ultra-high molecular weight polyethylene was reported to increase the strain energy density by about 150 % and the ductility by about 140 % (Dalton et al. 2003). Secondary crystallites were attributed an increase in the strain energy (Ruan et al. 2003). Similar behavior was observed in 1.8 vol% aligned MWCNT/polyacrylonitrile composites with an approximately 80 % increase in energy to yield and energy to break (Weisenberger et al. 2003). In addition, the Dalton research team also reported a process of spinning 60 wt% SWCNT/polyvinylalcohol fibers with pre-drawn energy absorbing capacity nearly 3.5 times spider silk (165 J/g), where the slippage between SWCNT bundles was suggested as the mechanism responsible for the enhancement in the toughness (Dalton et al. 2003).

Addition of 1 wt% MWCNT to polypropylene was reported to lead to an increase in crystallization rate of the composite. The modifications that happens in the morphology of a polymer matrix together with the debonding and pull-out energies required for CNTs, suggest that CNTs may augment the energy absorption or toughness characteristics of the composite (Assouline et al. 2003). A two-fold increase in the tension-tension fatigue strength was found for aligned SWCNT/epoxy composite, compared to typical carbon fiber/epoxy composites. Implanted CNT may efficiently prolong the formation of micro cracking and/or bridge crazing that can spread inside matrix and lead to fatigue failure (Ren et al. 2003).

Zoo added CNTs into *ultrahigh molecular weight polyethylene* (UHMWPE). The investigators concluded that the coefficient of friction increased slightly as the concentration of CNTs increased. They hypothesized that the improvement in the properties of CNT/UHMWPE composites was due to increased shear strength of the UHMWPE caused by addition of CNTs. The increase of friction coefficient, on the other hand, could be due to the change of morphology of the surface caused by CNT addition, as well as the increased shear strength (Zoo et al. 2004). Kanagaraj observed enhanced mechanical properties with an increase in CNT concentration, which is believed to be due to the good interface between polymer and CNTs, thus transferring the load from polymer to CNT (Kanagaraj et al. 2007). Mora also obtained the effective Young's modulus of an epoxy matrix reinforced with carbon nanotubes for different fiber volume fractions experimentally. They reported that for 14 and 17 % fiber volume fractions. The Young's modulus was obtained to be equal to 10 and 18.8 GPa, respectively, while the pure epoxy's Young's modulus was 1.2 GPa (Mora et al. 2009).

In a recent investigation, Liu team studied the CNT/Al nanocomposites. Metal matrices are capable of providing conductive composites; inclusion of reinforcing phase, however, reduces the conductivity of the material. The main goal of this study was hence to enhance the mechanical properties, while preserving its high electrical properties. For this purpose, powder metallurgy and subsequent friction stir processing techniques were mixed and utilized to fabricate uniformly dispersed carbon nanotube reinforced Al and 6061Al composites. Based on the results of this approach, the incorporation of CNT boosted the tensile properties of both the Al and 6061Al composites. However, although the electrical conductivity of the composites was lowered in case of CNT/Al, the existence of Mg and Si elements, employed for segregation at CNT/6061Al interface, leaded to an enhancement in electrical properties of the nanocomposites (Liu et al. 2014).

3.5.2 Electrical Properties

As explained earlier, single walled carbon nanotubes may behave either as a metal or as a semiconductor, depending on their chiral angle, but multi walled carbon nanotubes are always electrically conductive, with an electrical conductivity of approximately 1.85 kS/cm (Ando et al. 1999). On the other hand, CNTs in metal matrices are also showing high conductive characteristics. Hence, the electrical properties of CNT based composites have always been of major interest. However, when the amount of fillers increase in the composite, leading to higher conductivity, the mechanical properties of the composites lowers significantly. Therefore, many investigations focus on both properties together to obtain an optimum CNT content, like the investigation conducted by Uddin team to improve both hardness and electrical conductivity of CNT/Cu composites (Uddin et al. 2010).

3.5.2.1 Polymer and Ceramic Matrices

In one of initial investigations on the mechanical and electrical behavior of MWCNT-based nanocomposites, Allaoui team fabricated and tested multi-walled carbon nanotube/epoxy resin composites in 2002. They achieved a rubbery epoxy resin by choosing an over-aged relatively soft hardener, and a ductile matrix and studied the mechanical and electrical properties of the composite with different weight percentages of nanotubes. According to their investigations, by adding only a nanotube concentration between 0.5 and 1 wt% the composite transition from insulator to conductor took place (Allaoui et al. 2002). Later, Sandler team reported an ultra-low electrical percolation threshold in carbon-nanotube-epoxy composites at 0.0025 wt% of CVD-grown aligned MWCNTs. The low percolation threshold, which is an order of magnitude smaller than that with entangled multiwall nanotubes is believed to be due to alignment of CNTs. Sufficient conductivity for anti-static applications was obtained with an average nanotube content of approximately 0.005 wt% in epoxy matrices (Sandler et al. 2003).

Electrical characteristics of metal-semiconductor devices fabricated from composites of high molecular weight polyaniline and various weight percentages of SWCNTs were investigated. Current-voltage studies on these devices showed a remarkable increase in current as carbon nanotube concentration in the composite increased (Ramamurthy et al. 2004). In another study, the influence of CNT contents on electrical and rheological properties of CNT-reinforced polypropylene composites showed that the volume resistivity of the composites was decreased with increasing the CNT content and electrical percolation threshold was formed between 1 and 2 wt% CNT, due to the formation of conductive chains in the composites. The viscosity of the composites was increased with the addition of CNT, leading to an increase in elastic melt properties, which is a result of the higher aspect ratio of the CNT. In addition, composites containing more than 2 wt% CNT exhibited non-Newtonian curves at low frequency (Seo and Park 2004).

Considering that the common production techniques of CNTs result in defected or unpurified nanotubes, Kim investigates the effect of two common chemical purification techniques, i.e. oxidation by either nitric acid or by the mixture of H_2O_2 and NH_4OH solution, on the electrical properties of the final nanocomposite. According to their results, the latter solution yields epoxy composites with a higher conductivity than those produced with the MWCNT oxidized by nitric acid (Kim et al. 2005).

After Li team's successful approach in homogenous distribution of CNTs inside matrix (Li et al. 2005), Moisala and his research team used their technique to continue Sandler's investigation on electrical conductivity enhancement of MWCNT-epoxy composites at low filler concentration (Sandler et al. 2003) by using SWCNTs as the filler material. In their investigation, epoxy composites containing 0.005–0.5 wt% of both SWCNTs and MWCNTs have been studied. The MWCNT composites had an electrical percolation threshold of less than 0.005 wt%, whereas in the SWCNT composites the electrical percolation thresholds were higher in the range between 0.05 and 0.23 wt% (Moisala et al. 2006).

Although CNT/ceramic composites has been attractive materials for their enhanced mechanical properties, the possibility to control their electrical conductivity through changing the CNT content in them made them good options for electrical applications as well. A family of SWCNT-$MgAl_2O_4$ has been prepared, with a range of 0.23–24.5 vol% CNT content distributed homogeneously inside the matrix. A jump of the DC electrical conductivity has been measured and the electrical conductivity is well fitted to the scaling law of the percolation theory up to 11 vol% of CNT content (Rul et al. 2004). The variety of methods employed for producing ceramic and metal matrix nanotube, however, plays an important role in the properties of the resulted composite and its mechanical and electrical performance, depending on the choice of material, debonding and interfacial sliding characteristics (Samal and Bal 2008).

3.5.2.2 Metal Matrices

High thermal and electrical conductivity of metal matrices was a major motivation to migrate from polymer matrices to metals matrices, such as aluminum or copper. However, compared to polymer matrices, these structures were hard to produce in terms of uniform dispersion of fillers, good interface strength and bulk production (Esawi and Farag 2007). These three main issues are the major concerns in producing the next generation of metal/CNT nanocomposites (Kwon et al. 2010).

Starting since 1998, there has been several investigations on manufacturing carbon nanotube-aluminum composites and manipulating their properties towards better outcomes using different techniques, namely high-energy ball milling, cold isostatic pressing, hot extrusion, and a powder rolling technique (Kuzumaki et al. 1998; Esawi and Morsi 2007; Deng et al. 2007; Esawi and Borady 2008; Estili and Kawasaki 2008; Sridhar and Narayana 2009; Kwon et al. 2009). However, achieving homogenous dispersion and acceptable interface properties was still a matter of importance. Cu on the other hand had the potential for applications in various electronic devices, making it an excellent candidate as another metal matrix for producing CNT-based nanocomposite materials.

There were difficulties in producing homogenously dispersed CNTs within Cu matrices as well, especially at high CNTs loading and there was a very weak bonding between carbon nanotube and pure copper matrix. In a recent investigation, Chu and his research team successfully modified the Cu matrix by adding Ti to retain the electrical conductivity of the composite, while enhancing its interfacial bonding and yield strength (Chu et al. 2013a). Before, they suggested that by using the particles-compositing technique the problem of uniform distribution of CNTs inside Cu matrix could be solved (Chu et al. 2010). However, according to their investigations, an increased amount of CNT fibers leads to lowered electrical conductivity of the composites. It happens because the electrical conductivity of CNTs is lower than that of Cu, and hence, the electrical conductivity of the composites will degrade based on the rule of mixtures. Furthermore, porosity will happen inside the composite, making insulation sites and finally as CNTs always locate themselves at the grain boundaries of Cu matrix, boundary resistance for electron movement will increase significantly (Chu et al. 2013a).

3.5.3 Thermal Properties

Other than superior electrical and mechanical properties, CNTs are known to have one of the highest thermal conductivities and hence, they are attractive for being employed in composites for different thermal management purposes as well (Kim et al. 2001). The thermal conductivity of CNTs has been predicted to be up to 6000 W/mK (Esawi and Farag 2007; Baughman et al. 2002; Erik et al. 2001), therefore, many investigations have focused on manipulating thermal properties of CNT based composites. In addition, as electronic devices become slimmer, heat management become a key factor in device design and application. Similar concerns should be considered in some other applications, such as electric motors and generators, heat exchangers in power generation, and automotives (Han and Fina 2011). Therefore, many attempts have been made to use highly thermally conductive polymer based composites instead of metallic materials, because of their lightweight, corrosion resistance, easy processing and lower manufacturing costs. As these types of composites require thermally conductive fillers, CNTs became a promising candidate for the purpose.

3.5.3.1 Polymer Matrices

In an early investigation, CNT reinforced polymer composites were simulated using molecular dynamics, seeking potential applications of a novel material with thermal expansion and diffusion coefficients higher than glass transition temperature. This property leads to a significant increase in the thermal conductivity of CNT composites at high temperatures, which was expected to have implications in CNT composite processing, coating and painting application. Classical molecular dynamics was employed using Brenner potential for modeling intra-nanotube interactions and Van der Waals forces at polymer-nanotube interfaces. In this study, temperature dependent structural behavior of carbon nanotube-polyethylene composites was investigated. According to this initial investigation the thermal expansion coefficient of the composite and diffusivity of polymer molecules increases significantly above glass transition temperature of the composite, which was demonstrating a good starting point in employing CNT based composites for thermal management purposes (Wei 2002).

Later on, as mentioned earlier, Moisala team tried to complete Sandler team's investigations on the electrical properties of MWCNT based composites (Sandler et al. 2003; Moisala et al. 2006). However, they expanded the outcomes not only by studying SWCNT/epoxy composites, but by measuring the thermal properties of the produced composites as well. According to their results, the MWCNT composites had a modest thermal conductivity increase as a function of the filler content. In the case of the SWCNT composites, however, the thermal conductivity was lower than that of the pristine epoxy (Moisala et al. 2006).

In a numerical approach, Song and Youn used control volume FEM to study the effective thermal conductivity of the polymeric composites filled with carbon nanotubes by using the asymptotic expansion homogenization technique. The representative volume elements employed, including a single CNT each, are constructed by assuming that the CNTs are distributed homogeneously inside the polymer matrix. Based on the comparison they represented between the numerical results and experimental investigation, the homogenization method was validated as a promising tool for predicating effective thermal conductivity of CNT based composites, due to being capable of considering anisotropic material properties such as thermal conductivity, elastic modulus, and thermal expansion coefficient as well as geometrical complexity (Song and Youn 2006). In another approach, a theoretical model was also developed by Bagchi and Nomura in 2006 to predict the effective thermal conductivity of aligned multi-walled nanotube polymer composite. To model the composite, the effective medium theory was employed for composites containing aligned spheroid inclusions with imperfect interfaces and the geometry of CNTs were incorporated into this theory by a continuum model that considered its structure and the mechanism of heat conduction. They concluded that regardless of the high nanotube-matrix boundary resistance, the reason behind the lower overall conductivity compared to its expected value, is that in the composites, the outer nanotube layer carries the bulk of the heat flowing through the nanotube. Furthermore, the effective conductivity was proved to be highly sensitive to the nanotube diameter as well (Bagchi and Nomura 2006).

The effect of interface on thermal conductivity of the composite was studied using continuum mechanics approach. Based on this numerical analysis, the effect of interface on the overall thermal conductivity of composite was found to be very small for short nanotubes, whereas in case of long tubes, the interface had a significant effect on the overall conductivity. Furthermore, in longitudinal direction, the nanotube anisotropy did not have a remarkable influence on conductivity, but transverse thermal conductivity was changing linearly with nanotube length (Singh et al. 2007). Fiamegkou team recently employed FEM to study the effect of various carbon nanotubes weight fractions on the effective thermal conductivity of CNT-reinforced polymer composites. They investigated the influence of various CNT contents using the parametric study of the Kapitza resistance value and the thickness of the intermediate layer and compared the predicted results with experimental measurements of equivalent CNT/epoxy resin composites. The good agreement between the calculated and the measured values proves that in order to achieve accurate simulation results, the existence of the thermal resistance between the CNTs and the polymer matrix should be considered by associating Kapitza resistance (Fiamegkou et al. 2014).

3.5.3.2 Metal Matrices

As mentioned earlier conductivity is provided in composites either by having conductive fillers, like MWCNTs, or by employing conductive matrices. Hence, metal matrices with intrinsic high thermal and electrical conductivity have been investigated in this scope as well. Kim studied the thermal conductivity of CNTs dispersed inside Cu matrix. The nanocomposites were synthesized via the molecular-level mixing process and based in their report, although inclusion of CNTs improved the thermal conductivity of CNT/Cu nanocomposites, the conductivity was decreased by increasing the CNT volume fraction, due to the interfacial resistance between CNT and Cu (Kim et al. 2011). The results obtained by Cho team were showing better values for thermal conductivity. They employed hetero-aggregation method based on a wet mixing process to produce CNT/Cu composites. The thermal conductivities obtained for CNT/Cu nanocomposites were higher than that of unreinforced Cu up to 3 % CNT volume fraction. However, after reaching a pike of 359.2 W/mK, a descending trend was reported. For about 5 and 10 % of CNT volume fractions, the value of thermal conductivity was observed to be even lower than that of unreinforced Cu due to several reasons, including interfacial thermal resistance, porosities, inhomogeneous distribution and incomplete consolidation (Cho et al. 2010).

In an effort for controlling the uniform dispersion of MWCNTs inside Cu matrix, Chu team employed a novel fabrication technique. They used a new particles-compositing technique for dispersing MWCNTs inside Cu powder and consolidate it in form of CNT/Cu composites by spark plasma sintering (SPS) method. The introduced technique was capable of achieving homogeneous distribution of up to 10 vol% of MWCNTs. At higher CNT densities, clusters were appearing. Theoretical studies were also accomplished for comparison with available experimental findings of the effective thermal conductivity of the composites (Chu et al. 2010). Later on, the team tried to enhance the effective thermal conductivity, which turned out to be not very much fascinating in their previous investigation, by adding Ti to the Cu matrix and making CNT reinforced Cu–Ti alloy (CNT/Cu–Ti) composites. With this approach, the interfacial bonding of CNT/Cu–Ti composites is obviously improved by 7.5 and 15.1 %, at CNT loadings of 5 and 10 vol%, respectively. The enhancement is believed to be due to the formation of a thin TiC layer at the interface and hence a better interfacial bonding (Chu et al. 2013b).

Finally, in a recent investigation, since CNTs have much higher conductivity along their main tube axis, Afrooz et al. investigated the thermal conductivity of CNT/Cu nanocomposites using numerical analysis. The thermal conductivity of the nanocomposite models was predicted to improve compared to the unreinforced matrix. However, distribution of CNTs in the metal matrix played an important role in determining the thermal conductivity of nanocomposites, even when CNTs orientation and volume fraction were constant. As a result, two key factors was reported as requirements for achieving high thermal conductivity, namely homogeneous dispersion of CNTs inside the matrix with less agglomeration, and extremely low thermal resistance between CNTs and matrix (Afrooz et al. 2014).

3.5.4 *Optical Properties*

While still preserving enhanced mechanical properties, conjugated polymer composites can show advanced electrical and optical properties of semiconductors and metals (Bell et al. 2008). In addition, polymer composites have potential applications, in organic electronics, as the active component in various devices, like light-emitting diodes, photovoltaic cells and disposable electronic chips. However, an efficient contact between the matrix and the reinforcing phase is a prerequisite for efficient charge transfer in these composite materials (Star et al. 2004; Spanggaard and Krebs 2004). Therefore, composites of carbon nanotubes have been studied for being employed in photovoltaics, due to their high electron conductivity (Kymakis et al. 2003, 2006). Purified nanotubes have an extremely high surface area, making them good options for being employed as efficient exciton dissociation. However, due to some limitations, CNTs were as successful as expected in finding many optical applications outside the borders of research laboratories (Bell et al. 2008).

3.6 Defects Involved with CNT-Based Composites

When it comes with the study of nanomaterials, other than macroscopic defects, such as the quality of the dispersion of the fibers inside matrix, microscopic defects, such as the quality of adhesion between the fibers and the matrix, become very important too. In this section, some major defects are introduced as follows:

3.6.1 CNT Defects

The degree of solubility and the quality of dispersion of fiber inside the supporting matrix depend mostly on the fiber and matrix types and properties. SWCNTs, DWCNTs, or MWCNTs exhibit different properties in terms of chemical functionality along their basal-plane sidewalls and tip-ends. Moreover, shorter CNTs can disperse more effectively compared to longer CNTs (Bell et al. 2008). Various techniques are employed to boost the quality of dispersion of CNTs inside matrices; however, the effect of these treatments on the properties of the resultant composite is important to be considered. Reduction of electrical properties as a result of improving fiber-adhesion is a very common side effect of chemical treatment of CNT side-walls. Compared to theoretically defect-free CNTs, chemically treated CNTs that are kind of defective CNTs, which show different properties when incorporated as fibers inside matrices. Electronic properties of the composite are specifically more reliant on the structure of the CNTs to be as defect-free as possible.

However, other than direct chemical treatment of CNTs for better adhesion, the production procedure of carbon nanotubes generally involves with introduction of defects to the perfect structure. Various fabrication techniques often require high temperature synthesize in presence of metal catalysts, i.e. Fe, Mo, Ni or Co. As mentioned earlier chemical vapor deposition and arc-discharge are the two most commonly used techniques for production of CNTs, which both require a purification step to separate CNTs from amorphous, graphitic carbon and the metal catalyst. This step normally leads to CNTS being shortened and somewhat damaged. Hence, the products are exposed to high-temperature annealing to repair the damage to some extent. Therefore, careful selection of the production and purification process, as well as choosing an appropriate solvent system for casting a polymer composite, are key factors that plays a very important role in achieving acceptable properties for all types of carbon nanotubes in employed for electronic applications (Bell et al. 2008).

In case of MWCNTs, oxidation is one of the most commonly used methods of purifying that depending on its conditions, leads to damage in the structure of CNTs (Kim et al. 2005) and the opening of their caps (Feng et al. 2003) and subsequently influencing the properties of the resulted composites. For instance, with increasing the time of oxidation of CNT fibers in carbon fiber/polymer composites, their conductivity was reported to decrease, due to increase of the contact resistance between the fibers (Liang et al. 2000). In addition, the distance between neighboring conductive clusters that is dependent on the oxidation layer thickness, also influences the conductivity of the composite (Kim et al. 2005) and hence the chemical treatment and conditions opted for purifying of MWCNT fibers has direct influence on the physical and electrical properties of the composites.

3.6.2 CNT Bundles

Synthesis process of CNTs normally leads to variety in terms of lengths and diameters of the produces nanotubes, as well as creation of CNT bundles. Hence, the interaction between the supporting matrix and the reinforcing fibers is not clear and varies, creating a range of aggregation and crystallization behaviors (Bell et al. 2008). The nature of interfacial interaction between nanotubes and the matrix, therefore, depends not only on the matrix stiffness, but also on incorporated nanotube properties (Goh et al. 2006) and surface area (McCarthy et al. 2002). This variety of the produced clusters will subsequently influence the properties of the resulted composite; their conductivity for example. Therefore, different physical and chemical approaches have been utilized for separating CNTs from each other.

Carbon nanotubes have normally high aspect ratio that creates a non-covalent attractive interaction between them and subsequently aggregation of CNTs into bundles. Hence, in case of application's requirement, which is in fact a very common prerequisite of different applications, homogenous dispersion and distribution of separated CNTs would be necessary. For debundling of CNTs different approaches can be employed, like sonication in accordance with addition of a surfactant in solution with a proper solvent (O'Connell et al. 2001).

3.6.3 CNT Distribution

As mentioned earlier, fiber orientation and concentration play important roles on the properties of composites. At nanoscale however, analytical investigation on the influence of defects in the arrangement of CNTs inside matrix would be very difficult. On the other hand, experimental studies would be also very expensive. Hence, modeling and simulation of nanocomposites behavior under various situations are very likely to be employed for predicting the influence of imperfect dispersion of fibers into matrix. In some investigations, innovative techniques for simulating non-homogenous dispersion of CNTs, inside the representative volume element, as schematically illustrated in Fig. 3.5, have been achieved. Afrooz et al. employed truss elements in finite element to represent CNTs and studied the effect of non-homogenous dispersion of fibers, in terms of distribution density, on Young's modulus of nanocomposite (Afrooz et al. 2012).

In terms of orientation, on the other hand, the fibers can be parallel or randomly distributed with respect to each other. Different approaches have been employed to predict the elastic behavior of CNT based composites with aligned short fibers. However, less attention has been paid to study randomly distributed short fiber composites. In initial investigations on the topic, simple two-dimensional (Tandon and Weng 1986) and three-dimensional (Benveniste 1987) random arrangements of CNTs have been investigated. The general orientation distribution (Ferrari and Johnson 1989) was also studied briefly in 1989. Though, in

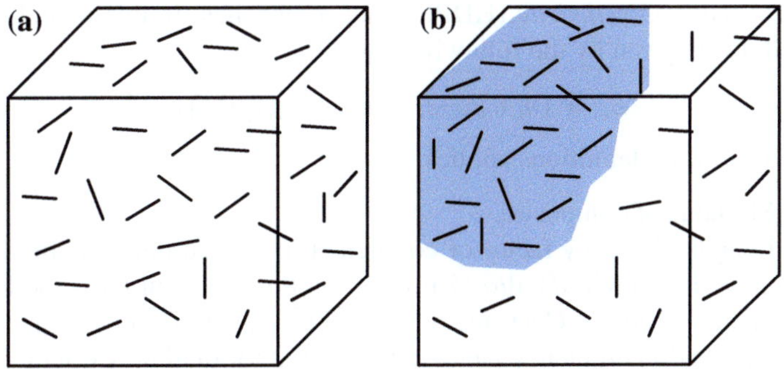

Fig. 3.5 Random distribution of CNT fibers inside a polymeric matrix, in terms of **a** orientation and **b** orientation and distribution density

more recent investigations for simulating various CNT orientations, introducing CNT/Matrix islands with various CNT degrees was suggested by Makvandi and Öchsner (2014) and randomness, in terms of distribution density, was also studied by Afrooz et al. (2012).

3.7 Analytical Studies

In order to predict the behavior of composites before experimentally fabricating them or for achieving the optimum characteristics for a specific application, analytical studies and computational modeling would be the best choices. However, analytical techniques are normally capable of predicting the properties of homogenous and perfect structures, whereas defects, that are inevitable parts of any material, are more likely to be introduced to the model using computational techniques.

3.7.1 Rule of Mixture

The most commonly used theoretical model for the analysis of elastic properties of CNT-reinforced composites is the *rule of mixture* (RM), which was modified by Coleman et al. (2006). While many other models are available, the RM is by far the simplest and the most intuitive one. In this model, it is assumed that both CNT and polymer are very well-bonded, equally strained and CNTs are homogeneously distributed within the polymer matrix.

In the theory of short fiber-reinforced composites, when fibers are randomly distributed, the composite modulus is related to the fiber modulus by the well-known mixing rule. Based on the two models, a *modified series RM model* is defined for randomly oriented CNTs distributed within the polymer matrix like

CNT-*high density polyethylene* (HDPE) nanocomposites. In this model, the composite modulus is given by the following equation:

$$E_c = (\eta_L \eta_o E_{CNT} - E_{PE})V_{CNT} + E_{PE}, \tag{3.10}$$

where the following definitions applies:

E_c : Modulus of composite,
η_L : Length efficiency factor or coefficient of critical fiber length; defined as $\eta_L = 1 - [\tanh(al/d)]/(al/d)$; where in this equation l/d is the length per diameter ratio of CNT and $a = \sqrt{(-3E_{PE})/(2E_{CNT}\ln(V_{CNT}))}$,
η_o : Orientation efficiency factor, which is 0.2 for randomly oriented fiber,
E_{CNT} : Modulus of CNTs,
E_{PE} : Modulus of polyethylene and
V_{CNT} : Volume fraction of CNTs

Wagner et al. showed that η_L, the critical fragmentation length for MWCNTs is less than 500 nm, which is always greater than the length of the tubes (Wagner et al. 1998). Therefore, the factor of length efficiency was set to be equal to one. Fisher et al. used a finite element analysis to demonstrate that even a slight curvature of CNTs significantly reduces the effective reinforcement (Fisher et al. 2003). Furthermore, the critical length of curved nanotubes could be much longer than that of the straight ones. Zhu and Nart, on the other hand, studied the effect of the nanotube orientation on the tensile modulus of carbon nanotube-reinforced polymer composites (Zhu and Nart 2004). Based on their numerical simulations using the finite element method, they found that the composite's Young's modulus decreases with increasing the orientation angle.

3.7.2 Halpin-Tsai Method

The observed differences in experimental and theoretical values show that reinforcement ability of the high Young's modulus of MWCNTs is not fully exploited in the modified series RM model. A possible explanation is the inability of the model to reflect the three-dimensional curving and coiling of MWNTs in a matrix. Therefore, another model was suggested by Halpin et al., which is known as the *Halpin-Tsai model* and is based on self-consistent micromechanics method (Halpin and Kardos 1976) and is described by the equation below:

$$\frac{E_C}{E_{PE}} = \frac{3}{8}\left[\frac{1 + 2(l/d)_{CNT}V_{CNT}}{1 - \eta_L V_{CNT}}\right] + \frac{5}{8}\left[\frac{1 + 2\eta_T V_{CNT}}{1 - \eta_T V_{CNT}}\right], \tag{3.11}$$

where

$$\eta_L = \frac{(E_{CNT}/E_{PE}) - 1}{(E_{CNT}/E_{PE}) + 2(l/d)_{CNT}}, \tag{3.12}$$

and

$$\eta_T = \frac{(E_{CNT}/E_{PE}) - 1}{(E_{CNT}/E_{PE}) + 2} \tag{3.13}$$

Later, Kanagaraj et al. concluded that by using the *Halpin-Tsai model* and *modified series RM model* together, Young's modulus of CNT-HDPE nanocomposites can be predicted with up to 2 and 4 % relative error in the concentration range from 0 to 1 wt% (Kanagaraj et al. 2007). Thus, a modified form of rule of mixture and the Halpin-Tsai equations are used nowadays to predict Young's modulus of CNT-HDPE composites as they consider both the *volume fraction* and *aspect ratio* of reinforcement to predict stiffness of the composites (Tucker and Liang 1999).

3.8 Modeling Techniques

With the aid of desktop computers, the most cost effective way of characterizing nanocomposites was achieved. Computational approaches can play important roles in predicting the behavior of CNT composites. Molecular dynamics simulations and continuum mechanics are the two most famous techniques to understand the material properties of the composites. MD simulations are used for small-scale models, whereas continuum mechanics is a typical approach used for large-scale simulations. The limitations of MD approach are because it requires a large amount of memory on the computer to solve large-scale problems. Continuum mechanics, on the other hand, can overcome this problem for characterizing large-scale models. FEM and BEM are the two most commonly used continuum mechanics techniques in characterizing nanocomposites.

3.8.1 CNT/Matrix Unit Cell

The nanocomposite structure of CNTs implemented inside a unit cell of polymer matrix was studied with the purpose of characterizing the molecular structure and physical and mechanical properties. The unit cell can be rectangular and cylindrical. It is interesting to know how the presence of a supporting polymer matrix, for instance, influences the mechanical properties of the whole structure. In Fig. 3.6, the structure of a sample CNT inside a (a) cylindrical and (b) rectangular cell of embracing matrix is illustrated.

As mentioned earlier, a single-walled carbon nanotube can be imagined as a graphene sheet that has been rolled into a tube. This tube was assumed to be equivalent to a hollow cylinder in the modeling process that has all the geometric and mechanical properties of the real carbon nanotube (Mavalizadeh et al. 2011). As mentioned before, although different values have been assigned for the

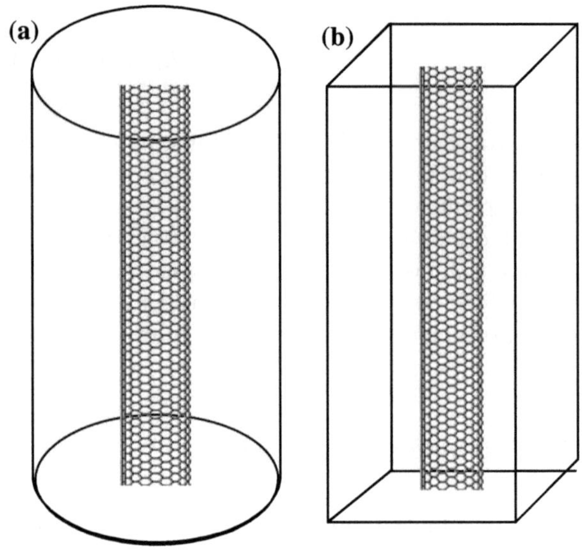

Fig. 3.6 An SWCNT held inside **a** cylindrical and **b** rectangular unit cell of the polymer

thickness of SWCNT, 0.34 nm was the most commonly used thickness which was applied in this research as well (Zhu and Nart 2004; Zhong et al. 2002).

Kalamkarov et al. applied plane elements to build the whole tube structure (Kalamkarov et al. 2007). In Mavalizadeh's approach too, the matrix was represented by a single solid 3-D element, while the tube was approximated by four 2-D plane elements around it. Figure 3.7 shows the RVE of the initial test, as well as the applied boundary conditions.

To achieve this goal perfectly the unit cell was modeled in two forms. First, the hollow cylinders were modeled inside a cubic matrix and the cylinder's properties were identical to that of an SWCNT and then a cylindrical supporting matrix was defined to obtain the mechanical properties of the composite structure's unit cell. The schematic view of the leading concept of defining a unit cell or a CNT/matrix island is presented in Fig. 3.8. Figure 3.9 shows the representative cubic and cylindrical unit cells, whereas the sizes of the tube's inner and outer diameters compared to the dimensions of the supporting matrix, gives the volume fraction of the composite. The volume fraction of a composite is in fact a key property of the structure, which has a significant influence on deciding its mechanical behavior.

In the study of the effect of variation in the matrix volume fraction and material properties, on the effective properties of RVE with hollow and filled tubes, the volume fraction of a matrix can be defined as:

$$\varphi = \frac{V_m}{V_m + V_c}, \tag{3.14}$$

where V_{m} and V_{c} are volumes of the matrix and carbon nanotube, respectively.

Based on the definition of the matrix volume fraction, the geometry for each unit cell with a particular volume fraction can be calculated. Referring to Fig. 3.9,

Fig. 3.7 Model structure of the composite; **a** Overview, and **b** boundary conditions

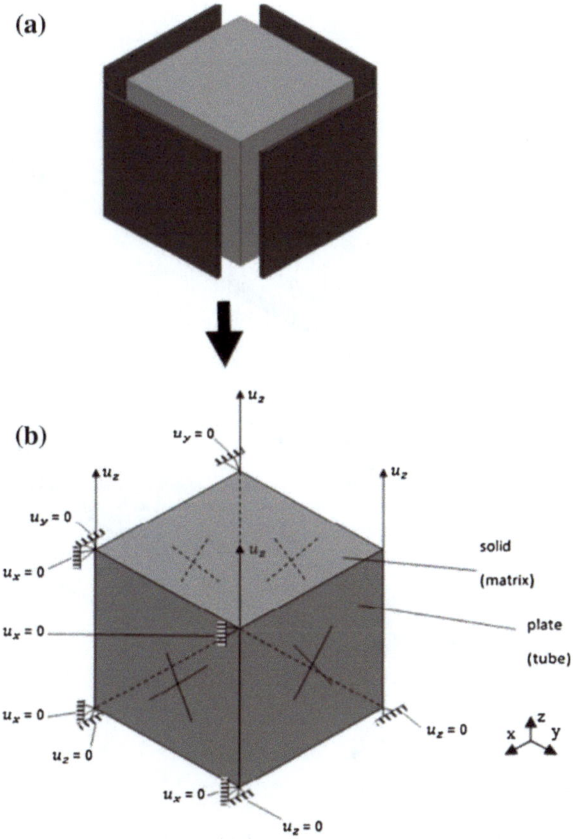

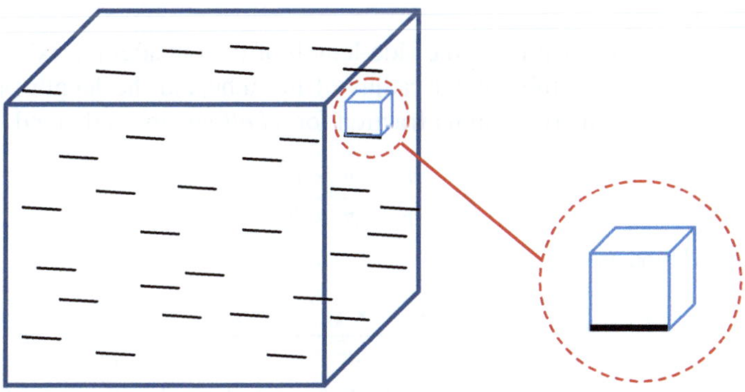

Fig. 3.8 A schematic view of a possible cubic unit cell out of a whole composite, i.e. reinforcement element and its supporting matrix

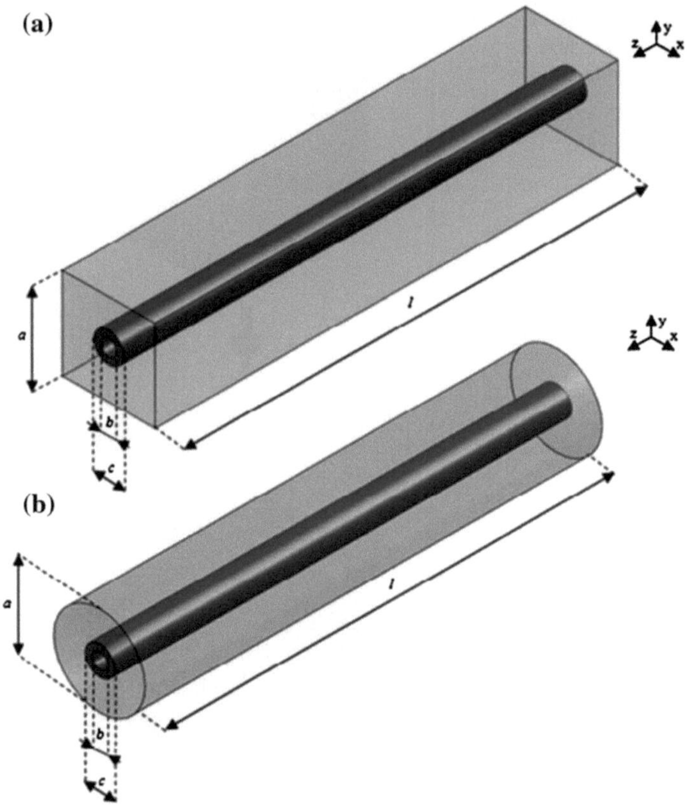

Fig. 3.9 Representative volume element of a SWCNT in microscopic matrix: **a** Cubic model, **b** Cylindrical model

if *a*, *b*, *c* and *l* are assumed to be the side-length of the outside-matrix's cross section, inner radius of the tube, outer radius of the tube,and the length of the unit cell, respectively, the matrix volume fraction for a *hollow* tube is defined as:

$$\varphi = \frac{a^2 - \pi \times c^2}{a^2 - \pi \times b^2}, \tag{3.15}$$

and for a *filled* tube as

$$\varphi = \frac{a^2 - \pi \times (c^2 - b^2)}{a^2}, \tag{3.16}$$

In the above equations, the length of RVE and the tube are assumed to be the same. Based on the mentioned equations, and by applying Hooke's law, i.e. $E = \sigma/\varepsilon$, in the elastic range, the highest value for the Young's modulus of a cubic unit cell was derived for the smallest matrix volume fraction and that it decreases approximately linear by increasing the matrix volume fraction for both *filled* and *hollow* tubes surrounded by matrix (Mavalizadeh et al. 2011). As one would expect, all Young's modulus values are between that of an individual tube and an individual matrix. Even for the filled tube structure *without* the surrounding matrix, the value of Young's modulus is in the same logical range. In addition, the value of the effective elastic modulus for the *filled* carbon nanotube is higher than for a *hollow* carbon nanotube and the difference between them decreases by increasing the matrix volume fraction, φ, to converge when the matrix volume fraction approaches a value of one. Figure 3.10 shows the applied boundary conditions on the cross-section of both cylindrical and cubic unit cells.

It is worth mentioning that in the calculations above, the effective cross section, A, was defined by:

$$A = A_{\mathrm{m}} + \sum A_{\mathrm{CNT}} \tag{3.17}$$

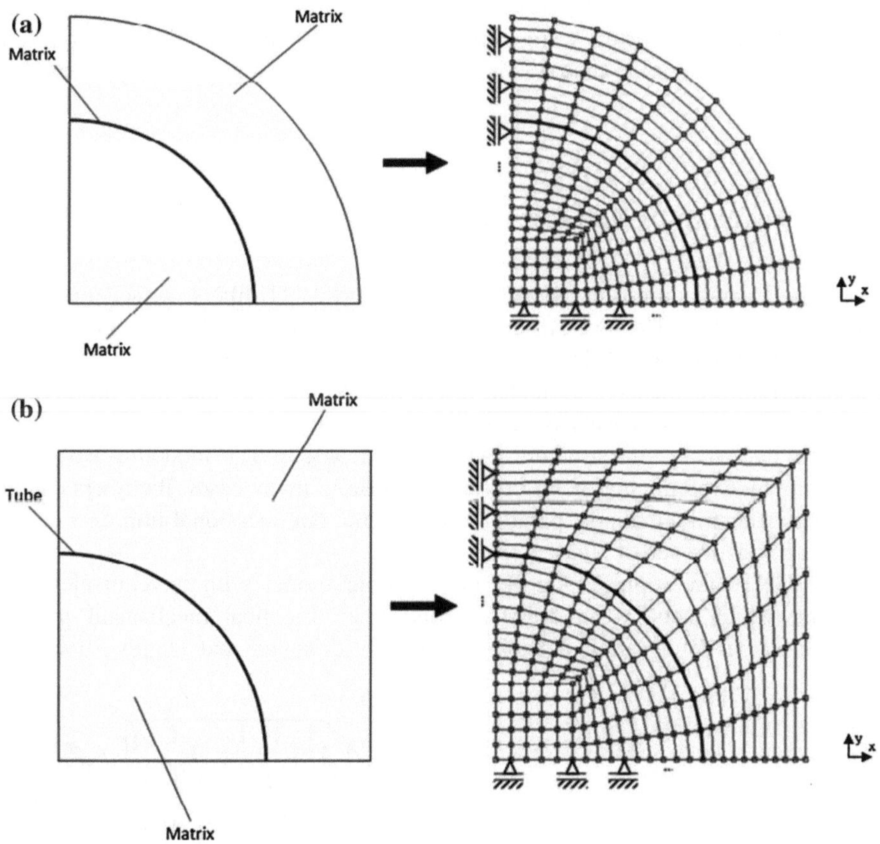

Fig. 3.10 One-fourth of the upper surface of RVE with applied FEM mesh and boundary conditions: **a** cylindrical, **b** cubic model

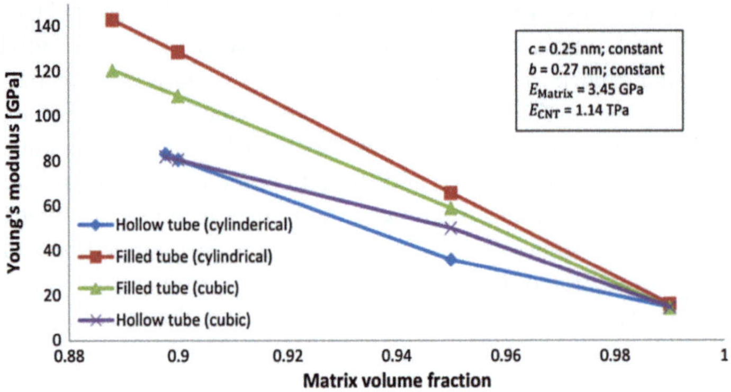

Fig. 3.11 Effect of matrix volume fraction on the longitudinal Young's modulus of cylindrical and cubic unit cells that contain filled and hollow CNT

where A_m and A_{CNT} are the effective cross sections of matrix and the CNTs respectively and the results were compared for filled and hollow CNTs supported by a cubic or cylindrical matrix. The results of this investigation are presented in Fig. 3.11.

3.8.2 *Modeling of Nanocomposite*

The whole nanocomposite, having parallel short CNT fibers, is schematically shown in Fig. 3.12. However, as explained earlier, modeling of non-homogeneous distribution of parallel CNTs was achieved by Afrooz et al. in 2012. In this study, the nano-reinforcements were added to the matrix in a way that they are parallel to each other but the density of reinforcement elements per matrix volume is not identical over the entire structure (Afrooz et al. 2012). The modeling was valuable from an experimental point of view, because in many cases, the fibers can be arranged to distribute almost parallel to each other but occasional improper segregation and occurrence of bundles are inevitable.

The CNTs are replaced in the macroscopic model with an equivalent FE element, which acts as an effective solid fiber. Identical mechanical properties, i.e. same deformation behavior, thickness, diameter and length, should be

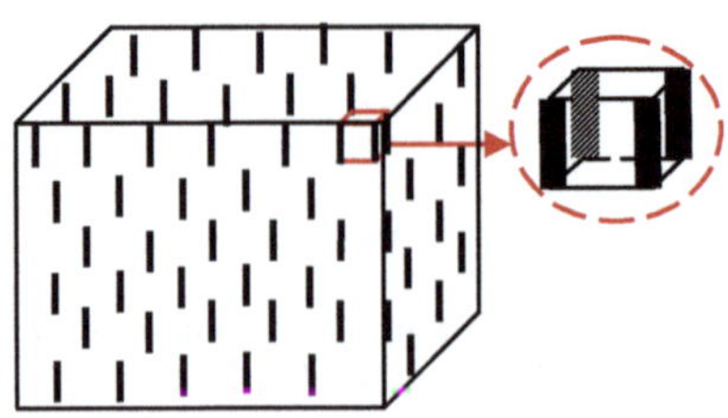

Fig. 3.12 Distribution of parallel CNTs inside a polymer matrix

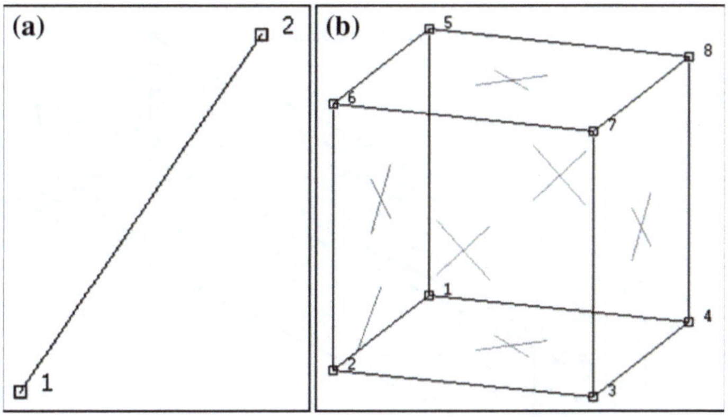

Fig. 3.13 **a** 3-D truss element; to replace CNTs, and **b** 3-D arbitrarily distorted brick element, to replace matrix

assigned to the 3-D truss element, representing the CNTs. On the other hand, as shown in Fig. 3.13, an eight-node, isoparametric, arbitrary hexahedral element, which has three degrees of freedom (translations only) per node, is opted to replace the matrix, with isotropic material properties. The stiffness of this element is formed using eight-point Gaussian integration. Figure 3.14 shows a unit

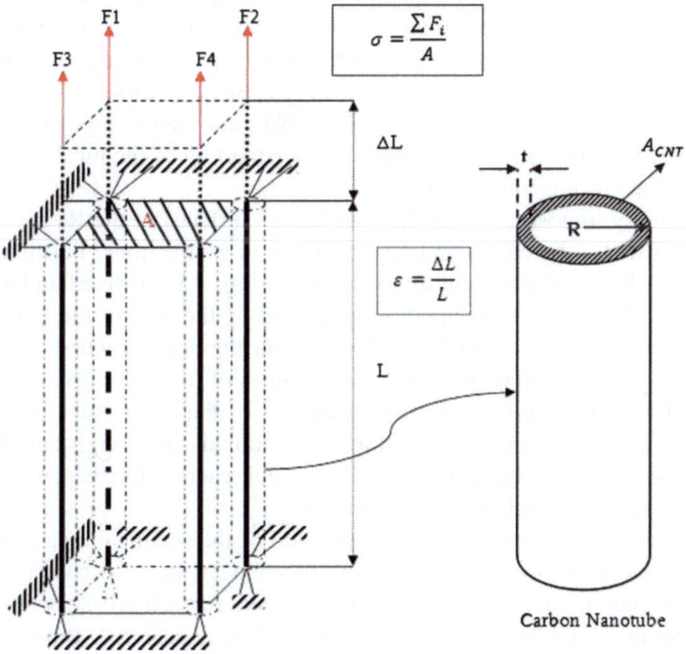

Fig. 3.14 Representative of a unit cell after replacing CNTs with truss elements. Reaction forces, cross-sectional area, displacement and initial length are defined by F_i (i = 1, 2, 3, 4), A, ΔL and L, respectively

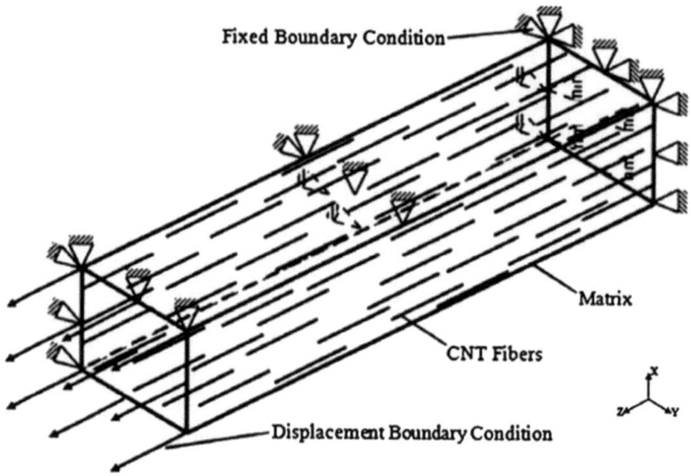

Fig. 3.15 Schematic representative volume element consisting of aligned CNT fibers with applied boundary conditions

cell of nanocomposite under tensile test in which the CNTs are replaced by truss elements.

After choosing the best corresponding elements to replace the matrix and the reinforcement fiber, the macroscopic behavior of the whole composite can be investigated. The generation of computational models, will help to stimulate the response of the perfect and imperfect nanostructures to different applied loads. Simulation of the deformation behavior of CNTs and nanocomposites, by means of compression tests, will lead to a better understanding of the mechanical properties of these advanced nanostructures.

As symmetry imposes, two of the four lateral surfaces of the RVE can be constrained in x and y directions and free to move in z directions. Furthermore, one end of the RVEs, i.e. the area of the right end in Fig. 3.15, is fixed and a displacement ΔL parallel to the fiber direction is applied on the area of the left end of the RVEs. The reaction forces will be calculated on the surface where the displacement boundary condition is prescribed. The values of these reaction forces will be employed for evaluating the effective properties of the composites. The effective Young's modulus of the composite, in the longitudinal direction, is calculated then by the Hooke's law, by obtaining the sum of nodal reaction forces generated on the surface on which the displacement is applied.

The total sectional area is defined as the sum of the sectional area of the matrix and the sectional areas of the fibers that are included in it. In case of randomly disturbed aligned fibers, since the number of fibers in each cross-sectional area is different, an averaged total sectional area should be considered. Based on several preliminary investigations, average area based on 11 sectional areas, as described below, would be sufficient to achieve acceptable representative values (Afrooz et al. 2012).

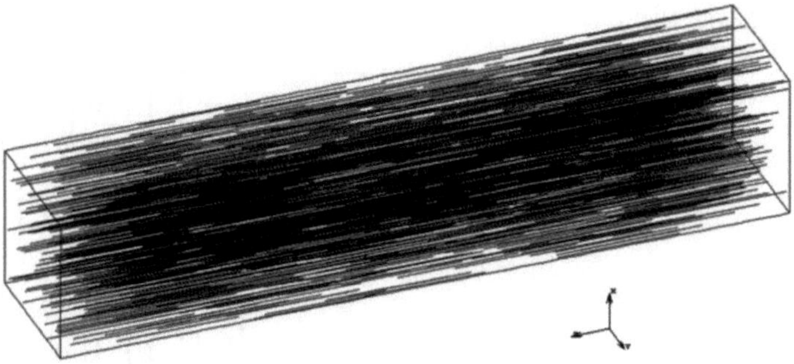

Fig. 3.16 Nanocomposite RVE model, containing 2000 randomly distributed CNT fibers (CNT length = 50 nm, volume fraction = 10.48 %)

$$A = \frac{\sum_{i=1}^{11} A_i}{11}. \tag{3.18}$$

However, regardless of the basic facts and equations, the method of distributing CNT fibers into the matrix model is the most crucial part of a modeling investigation. An RVE with 2000 randomly distributed CNTs is shown as an example in Fig. 3.16.

3.8.2.1 Modeling of Non-homogeneous Filler Dispersion

Composites with non-homogeneously dispersed fillers are very common to be achieved in experimental studies. Therefore, it is very much important to study the influence of random distribution of unidirectional CNT fibers on the mechanical properties of composites. In order to achieve this purpose, unidirectional CNT fibers of lengths 50 nm were used by Afrooz et al. for modeling large-scale composite with CNTs distributed randomly inside the matrix. Figure 3.17 is a schematic presentation of this model. To achieve this goal, various volume fractions were developed by increasing the number of CNTs inside the matrix, while holding the dimensions of CNT and matrix constant. The RVE was divided into smaller elements by using different mesh densities to generate various possibilities of places (nodes), where CNT fibers were placed between them. Randomness was defined as permitting various numbers of nodes in matrix for each RVE with constant number of fibers (Afrooz et al. 2012).

Since the dimensions of the matrix and the carbon nanotube are quite different, an algorithm was derived to automatically disperse the CNTs in the matrix and generate appropriate computational meshes with maximum number of opportunities for the fibers to implement inside their supporting matrix. Therefore, the RVE was repeatedly divided into smaller elements by using different mesh densities to generate a variety of possibilities of places (nodes) in which CNT fibers could be placed.

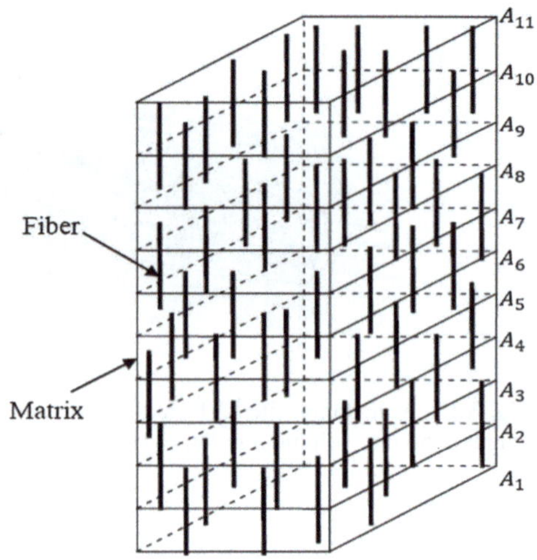

Fig. 3.17 RVE with 10 layers and 11 sectional areas

The density of nodes in the RVE is a decisive factor of randomness, meaning that as the number of nodes in the RVE increase, the number of possible places that could be occupied by a certain CNT will increase consequently. Hence, the more the potential fiber initiation points are, the more the extent of randomness will be. Therefore, CNT fibers (e.g. 2000 fibers) were distributed into the matrix in a way to avoid any contact between them. Young's modulus of the RVE composites with different random distribution of CNT fibers, obtained from the finite element simulations, is shown in Table 3.1 and compared with the results obtained by the Halpin-Tsai equation.

According to the results of this investigation, an increase in fiber volume fraction results in an increase of Young's modulus. Furthermore, the extent of randomness seems to affect the value of the obtained Young's modulus for a fixed volume fraction and to lead to more accurate results, compared to the result obtained from

Table 3.1 Young's modulus obtained for various volume fractions and distributions of fibers inside matrix

Fiber volume fraction (%)	Effective young's modulus (GPa)					
	29,791 nodes	58,621 nodes	102,541 nodes	226,981 nodes	400,221 nodes	Halpin-Tsai
	Randomness →					
0.00	3.8	3.8	3.8	3.8	3.8	3.8
2.50	10.49	8.74	8.24	7.71	7.32	10.86
5.00	16.50	13.11	11.92	11.08	10.32	18.28
10.48	28.80	21.87	19.35	17.51	16.10	36.02

analytical calculations. In other words, by decreasing the number of sites (nodes), which could be occupied by fillers in a given matrix, Young's modulus increases. Indeed, fillers have many possibilities for being distributed into the matrix during the process of experimental manufacturing. Therefore, there can be various cases in sense of distribution of fibers in a matrix. It is clear that the variation of the randomness factor has a significant effect on the obtained Young's modulus. It was also concluded that a model with more aligned fibers, i.e. less randomness, exhibits a convergence toward the Halpin-Tsai curve, proposed for aligned short-fiber composites (Afrooz et al. 2012).

The domain of values, which can be predicted for Young's modulus versus the volume fraction, is obtained as the gray area shown in Fig. 3.11. Shokrieh and Rafiee investigated the effect of the length of nanotube on the reinforcement efficiency in carbon nanotube based composites (Shokrieh and Rafiee 2010). Their result for the stiffness, which is obtained for a fiber length of 50 nm, is approximately 12.4 GPa. This value lies in the range of values which were predicted for the same length, i.e. 50 nm and fiber volume fraction of 5 % but different randomness (see the red region in Fig. 3.18).

Figure 3.19 is a plot of the estimated Young's modulus as a function of the extent of randomness, or in other words as a function of the number of available nodes for occupation, for different fiber volume fractions. The diagram clearly shows that the type of distribution of CNT fibers in the matrix, in terms of randomness, affects the Young's modulus. By holding the volume fraction constant, composites with more regular distribution of the fibers in the matrix, show a higher stiffness.

The effect of randomness becomes more significant at higher volume fractions. Tables 3.2 and 3.3 show the results of obtained Poisson's ratios, of ν_{zx} and ν_{zy}, respectively, for different fiber distributions in the matrix. In order to find out about the direction of X, Y and Z coordinates with respect to the element's direction you may refer to Fig. 3.15.

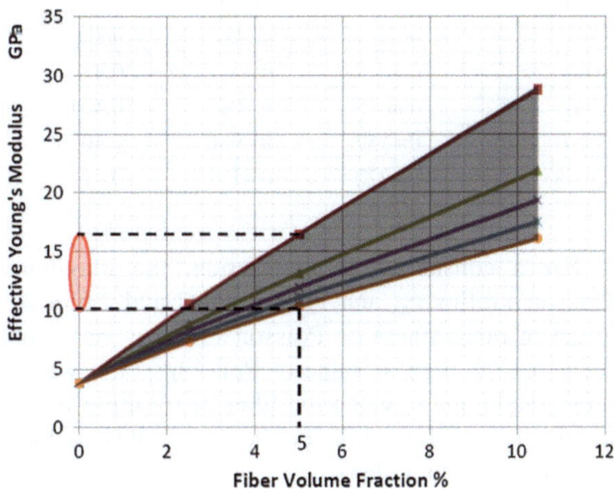

Fig. 3.18 The region of predicted Young's modulus values for 5 % fiber volume fraction

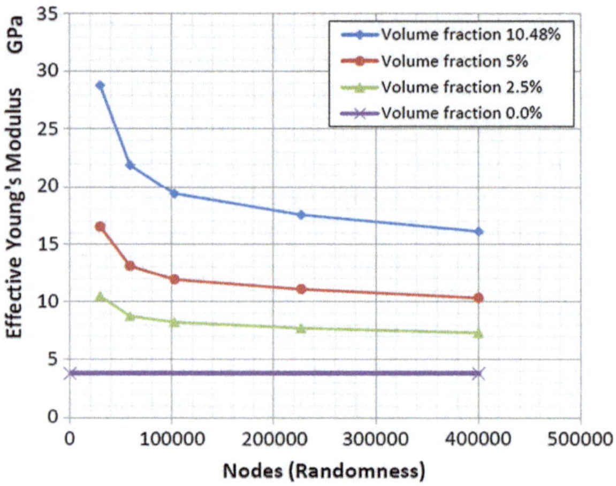

Fig. 3.19 Estimated longitudinal Young's modulus (GPa) versus number of nodes (*randomness*) in matrix for different fiber volume fraction

Table 3.2 Obtained Poisson's ratio of ν_{zx} for various volume fractions and distribution of fibers into the matrix

Fiber volume fraction (%)	Poisson's ratio (ν_{zx})				
	29,791 nodes	58,621 nodes	102,541 nodes	226,981 nodes	400,221 nodes
0.00	0.4	0.4	0.4	0.4	0.4
2.50	0.4091	0.4051	0.4044	0.4038	0.4036
5.00	0.4148	0.4085	0.4079	0.4066	0.4066
10.48	0.4237	0.4132	0.4125	0.4112	0.4110

Table 3.3 Obtained Poisson's ratio of ν_{zy} for various volume fractions and distribution of fibers into the matrix

Fiber volume fraction (%)	Poisson's ratio (ν_{zy})				
	29,791 nodes	58,621 nodes	102,541 nodes	226,981 nodes	400,221 nodes
0.00	0.4	0.4	0.4	0.4	0.4
2.50	0.4098	0.4043	0.4047	0.4037	0.4036
5.00	0.4157	0.4080	0.4078	0.4064	0.4067
10.48	0.4222	0.4121	0.4125	0.4106	0.4110

For a constant volume fraction, the maximum changes in the values of Poisson's ratios ν_{zx} and ν_{zy} are 2.997 and 2.653 %, respectively. As a result, the effect of randomness on Poisson's ratio is very small. However, it can be clearly seen that the highest value of Poisson's ratio was achieved in the range of minimum randomness, when the fiber distribution is more regular. It can be assumed that in case of aligned fibers, the composite can contract much easier than the case with more random fibers.

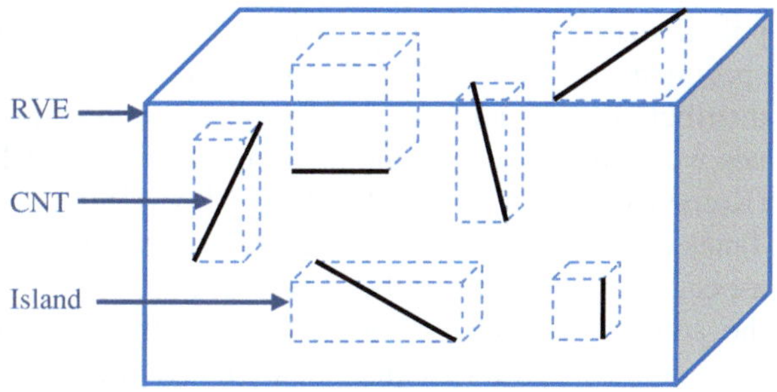

Fig. 3.20 Randomly-oriented CNT islands

3.8.2.2 Modeling of Non-parallel Filler Dispersion

In a more recent investigation, a new flexible and reliable approach was introduced for studying the mechanical properties of randomly oriented fiber composites, using the finite element method. Unlike previous studies in which a regular mesh was defined within a representative volume element, then the fiber randomly chose one of the mesh nodes for its position (Afrooz et al. 2012; Makvandi and Öchsne 2013), the main idea of the new technique is to create regular shaped islands around the distributed fibers and then to connect these islands to each other, and to the representative volume element by a proper meshing (Makvandi and Öchsner 2014). Figure 3.20 shows how an island is defined around each CNT fiber schematically. The new technique is capable of handling both aligned and randomly oriented arrangements of nanotube fibers.

3.8.3 Comparison of Experimental and Computational Studies

Several attempts have been devoted to compare the results that are achieved by experimental studies with what the theoretical investigations and computational modeling can predict. The topic is important because predicting the behavior of composites reliably will provide the chance of saving energy and cost of excessive experimental studies. Song and Youn compared the results they obtained for characterizing the effective elastic modulus of a carbon nanotube-based composite experimentally with analytical results (Song and Youn 2006). They used a control volume finite element, analytical solution. The results show that the numerically calculated elastic Young's modulus is in good agreement with that obtained by experimental investigation. They found Young's modulus to increase when the CNTs' aspect ratio and volume fraction raise and the trend is almost linear in the latter case.

Gómez-del Rio et al. studied nanocomposites based on *poly ethylene terephthalate* (PET) with small contents of SWCNT, which were prepared by in situ polymerization. Elastic constants were determined by tensile tests performed on specimens instrumented with strain gauges. They assumed random orientation distribution of nanotubes, experimental Young's modulus and Poisson's ratio values were compared to some micromechanical models, namely Cox and Krenchel (Krenchel 1964; Cox 1952), Halpin-Tsai (Halpin and Kardos 1976) and Mori-Tanaka (Mori and Tanaka 1973; Benveniste 1987), which took orientation and aspect ratio of the nanotubes into account. They found from the tensile tests that the addition of SWCNT in PET makes the nanocomposite stiffer (higher elastic modulus), harder (higher yield stress) and more brittle (lower failure strain) (Gómez-del Rio et al. 2010).

According to the results, the unreinforced matrix introduces a rougher fracture surface, i.e. a characteristic of a ductile material. However, by addition of SWCNT a much smoother fracture surface with some signs of brittle fracture will be achieved. This observation is in agreement with the values of yield stress and failure strain, obtained numerically. In other words, by increasing the nanotubes' volume fraction, the yield stress increases, but the maximum failure strain decreased.

Furthermore, the results show that by adding 0.3 % CNTs, the yield stress and Young's modulus increased 6.5 and 11.9 %, while the failure strain, on the other hand, decreases by 89.1 %. Although the failure strain of the structure drops dramatically, the better yield stress and Young's modulus of the new structure are still of significance importance in many devices that have vibrating motion, especially in elastic range. Poisson's ratio decreased by 3.2 % for the highest SWCNT content added.

Once material processing techniques are developed for good dispersion of CNTs in polymer matrices and modeling techniques are perfected and optimized, many potential material improvements and application could be investigated. There are strong industrial attempts in the synthesis, and process of more perfected inclusion of carbon nanotubes inside polymeric matrices for making advanced nanocomposites. The goal is to introduce these highly advanced composites in series of applications that are expected to lead the world of nano-bio-science and -technology toward many outstanding achievements. Hence, the exact and reliable prediction of the physical properties of these nanomaterials is very critical; which still requires more accurate investigations and applying new approaches.

References

Aboudi, J.: Mechanics of composite materials: a unified micromechanical approach. Stud. Appl. Mech. **29**, 338 (1991)

Afrooz, I.E., Öchsner, A., Rahmandoust, M.: Effects of the carbon nanotube distribution on the macroscopic stiffness of composite materials. Comp. Mater. Sci. **51**, 422–429 (2012)

Afrooz, I.E., Yusoff, P.S.M.B.M., Ahmad, F., Muhsan, A.S.: A numerical analysis for predicting the thermal conductivity of carbon nanotube reinforced copper-matrix nanocomposites. MATEC Web Conf. **13**, 04011-1–04011-5 (2014)

Ajayan, P.M.: Nanotubes from carbon. Chem. Rev. **99**, 1787–1799 (1999)
Ajayan, P.M., Iijima, S.: Capillarity-induced filling of carbon nanotubes. Nature **361**, 333–334 (1993)
Ajayan, P.M., et al.: Opening carbon nanotubes with oxygen and implications of filling. Nature **362**, 522–525 (1993)
Ajayan, P.M., Stephan, O., Colliex, C., Trauth, D.: Aligned carbon nanotube arrays formed by cutting a polymer resin nanotube composite. Science **256**, 1212–1214 (1994)
Allaoui, A., Baia, S., Chengb, H.M., Bai, J.B.: Mechanical and electrical properties of a MWNT/epoxy composite. Compos. Sci. Technol. **62**, 1993–1998 (2002)
Ando, Y., et al.: Physical properties of multiwalled carbon nanotubes. Int. J. Inorg. Mater. **1**, 77–82 (1999)
Assouline, E., et al.: Nucleation ability of multiwall carbon nanotubes in polypropylene composites. J. Polym. Sci. Part B: Polym. Phys. **41**(5), 520–527 (2003)
Bagchi, A., Nomura, S.: On the effective thermal conductivity of carbon nanotube reinforced polymer composites. Compos. Sci. Technol. **66**, 703–712 (2006)
Baughman, R.H., Zakhidov, A.A., De Heer, W.A.: Carbon nanotubes-the route toward application. Science **197**, 787–792 (2002)
Bell, J.M., et al.: Polymer-carbon nanotube composites: basic science and applications. In: Cairney, J.M., Ringer, S.P., Wuhrer, R. (eds.) Materials Forum, pp. 144–152. Institute of Materials Engineering Australia, Brisbane (2008)
Benveniste, Y.: A new approach to the application of Mori-Tanaka's theory in composite materials. Mech. Mate. **6**(2), 147–157 (1987)
Bianco, A., Kostarelos, K., Partidos, C.D., Prato, M.: Biomedical applications of functionalised carbon nanotubes. Chem. Commun. **5**, 571–577 (2005)
Bubert, H., et al.: Characterization of the uppermost layer of plasma-treated carbon nanotubes. Diam. Relat. Mater. **12**, 811–815 (2003)
Callister, W.D., Rethwisch, D.G.: Materials science and engineering: an introduction, 8th edn. Wiley, New York (2009)
Chen, C.H., Cheng, C.H.: Effective elastic moduli of misoriented short-fiber composites. Solids Struct. **33**(17), 2519–2539 (1996)
Chen, X.L., Liu, Y.J.: Square representative volume elements for evaluating the effective material properties of carbon nanotube-based composites. Comp. Mater. Sci. **29**, 1–11 (2004)
Chiang, C.R.: A statistical theory of the tensile strength of short-fiber-reinforced composites. Compos. Sci. Technol. **50**(4), 479–482 (1994)
Cho, S., et al.: Multiwalled carbon nanotubes as a contributing reinforcement phase for the improvement of thermal conductivity in copper matrix composites. Scripta Mater. **63**, 375–378 (2010)
Chu, K., Jia, C.-C., Li, W.-S., Wang, P.: Mechanical and electrical properties of carbon-nanotube-reinforced Cu-Ti alloy matrix composites. Phys. Status Solidi A **210**(3), 594–599 (2013a)
Chu, K., Jia, C.-C., Li, W.-S.: Thermal conductivity enhancement in carbon nanotube/Cu-Ti composites. Appl. Phys. A **110**, 269–273 (2013b)
Chu, K., et al.: Fabrication and effective thermal conductivity of multi-walled carbon nanotubes reinforced Cu matrix composites for heat sink applications. Compos. Sci. Technol. **70**, 298–304 (2010)
Coleman, J.N., Khan, U., Blau, W.J., Gun'ko, Y.K.: Small but strong-a review of the mechanical properties of carbon nanotube-polymer composites. Carbon **44**, 1624–1652 (2006)
Cox, H.L.: The elasticity and strength of paper and other fibrous materials. Brit. J. Appl. Phys. **3**, 72–79 (1952)
Dalton, A.B., et al.: Super-tough carbon-nanotube fibers. Nature **423**, 703 (2003)
Deng, C., Zhang, X., Ma, Y., Wang, D.: Fabrication of aluminum matrix composite reinforced with carbon nanotubes. Rare Met. **26**, 450–455 (2007)
Dresselhaus, M.S., Dresselhaus, G., Avouris, P.: Carbon nanotubes synthesis, structure, properties and applications. Springer, Berlin (2001)

Edidin, A.A., Kurtz, S.M.: Development and validation of the small punch test for UHMWPE used in total joint replacements. Key Eng. Mater. **198**, 1–40 (2001)

Erik, T.T., Ren, Z., Chou, T.W.: Advances in the science and technology of carbon nanotubes and their composites: a review. Compos. Sci. Technol. **61**, 1899–1912 (2001)

Esawi, A., Borady, M.: Carbon nanotube-reinforced aluminium strips. Compos. Sci. Technol. **68**, 486–492 (2008)

Esawi, A.M.K., Farag, M.M.: Carbon nanotubes reinforced composites: potential and current challenges. Mater. Design. **28**, 2394–2401 (2007)

Esawi, A.M.K., Morsi, K.: Dispersion of carbon nanotubes (CNT) in aluminium powder. Compos. Part A: Appl. Sci. Manuf. **38**, 646–650 (2007)

Estili, M., Kawasaki, A.: An approach to mass-producing individually aluminadecorated multi-walled carbon nanotubes with optimized and controlled compositions. Scr. Mater. **58**, 906–909 (2008)

Farsadi, M., Öchsner, A., Rahmandoust, M.: Numerical investigation of composite materials reinforced with waved carbon nanotubes. J. Compos. Mater. **47**, 1425–1434 (2013)

Feng, Y., et al.: Removal of some impurities from carbon nanotubes. Chem. Phys. Lett. **375**, 645–648 (2003)

Fereidoon, A., Saeedi, E., Ahmadimoghadam, B.: Comparison between different finite element methods for foreseeing the elastic properties of carbon nanotube reinforced epoxy resin composite. In: Proceedings of the World Congress on Engineering—Vol II, WCE-2008, London (2008)

Ferrari, M., Johnson, G.C.: The effective elasticities of short fiber composites with arbitrary orientation distribution. Mech. Mater. **8**(1), 67–73 (1989)

Fiamegkou, E., Athanasopoulos, N., Kostopoulos, V.: Prediction of the effective thermal conductivity of carbon nanotube-reinforced polymer systems. Polym. Compos. **35**, 1997–2009 (2014)

Fisher, F.T., Bradshaw, R.D., Brinson, L.C.: Fiber waviness in nanotube-reinforced polymer composites-I: modulus predictions using effective nanotube properties. Compos. Sci. Technol. **63**, 1689–1703 (2003)

Fukuda, H., Kawata, K.: On young's modulus of short fiber composites. Fiber Sci. Technol. **7**(3), 207–222 (1974)

Goh, R.G.S., Motta, N., Bell, J.I.M., Waclawik, E.R.: Effects of substrate curvature on the adsorption of poly(3-hexylthiophene) on single-walled carbon nanotubes. Appl. Phys. Lett. **88**(5), 053101–053103 (2006)

Gómez-del Rio, T., et al.: Influence of single-walled carbon nanotubes on the effective elastic constants of poly (ethylene terephthalate). Compos. Sci. Technol. **70**, 284–29 (2010)

Goze, C., et al.: Elastic and mechanical properties of carbon nanotubes. Synth. Met. **103**, 2500–2501 (1999)

Gupta, A.K., Harsha, S.P.: Analysis of mechanical properties of carbon nanotube reinforced polymer composites using continuum mechanics approach. Procedia Mater. Sci. **6**, 18–25 (2014)

Haitao, W., Zhenhan, Y.: Large scale analysis of mechanical properties in 3D fiber-reinforced composites using a new fast multipole boundary element method. Tsinghua Sci. Technol. **12**(5), 554–561 (2007)

Halpin, J.C., Kardos, J.L.: The Halpin-Tsai equations: A review. Polym. Eng. Sci. **16**, 344–352 (1976)

Han, Z., Fina, A.: Thermal conductivity of carbon nanotubes and their polymer nanocomposites: a review. Prog. Polym. Sci. **36**, 914–944 (2011)

Hashin, Z.: Analysis of composite materials. J. Appl. Mech. **50**(3), 481–506 (1983)

Hill, R.: Elastic properties of reinforced solids: some theoretical principles. J. Mech. Phys. Solids **11**(5), 357–372 (1963)

Iijima, S.: Carbon nanotubes: past, present and future. Phys. B **323**, 1–5 (2002)

Kalamkarov, A.L., Askari, D., Veedu, V.P., Ghasemi-Nejhad, M.N.: Generally cylindrical orthotropic constitutive properties modeling of matrix-filled single-walled nanotubes: axial mechanicalproperties. J. Compos. Mater. **41**, 757–779 (2007)

Kanagaraj, S., et al.: Mechanical properties of high density polyethylene/carbon nanotube composites. Compos. Sci. Technol. **67**, 3071–3077 (2007)

Kaw, A.K.: Mechanics of composite materials, 2nd edn. CRC Press Taylor and Francis, Boca Raton (2006)

Khare, R., Bose, S.: Carbon nanotube based composites- a review. J. Minerals Mater. Charact. Eng. **4**, 31–46 (2005)

Kim, P., Shi, L., Majumdar, A., McEuen, P.L.: Thermal transport measurements of individual multiwalled nanotubes. Phys. Rev. Lett. **87**(21), 215502-1-4 (2001)

Kim, Y.J., et al.: Electrical conductivity of chemically modified multiwalled carbon nanotube/epoxy composites. Carbon **43**, 23–30 (2005)

Kim, K.T., et al.: Influence of embedded-carbon nanotubes on the thermal properties of copper matrix nanocomposites processed by molecular-level mixing. Scripta Mater. **64**, 181–184 (2011)

Krenchel, H.: Fiber Reinforcement. Akademisk Forlag, Copenhagen (1964)

Kurtz, S.M., et al.: Validation of a small punch testing technique to characterize the mechanical behavior of ultra-high-molecular-polyethylene. Biomaterials **18**, 1659–1663 (1997)

Kuzumaki, T., Miyazawa, K., Ichinose, H., Ito, K.: Processing of carbon nanotube reinforced aluminum composite. J. Mater. Res. **9**, 2445–2449 (1998)

Kwon, H., et al.: Combination of hot extrusion and spark plasma sintering for producing carbon nanotube reinforced aluminum matrix composites. Carbon **47**, 570–577 (2009)

Kwon, H., Park, D.H., Silvain, J.F., Kawasaki, A.: Investigation of carbon nanotube reinforced aluminum matrix composite materials. Compos. Sci. Technol. **70**, 546–550 (2010)

Kymakis, E., Alexandrou, I., Amaratunga, G.A.J.: High open-circuit voltage photovoltaic devices from carbon-nanotube-polymer composites. J. Appl. Phys. **93**, 1764–1768 (2003)

Kymakis, E., Koudoumas, E., Franghiadakis, I., Amaratunga, G.A.J.: Post-fabrication annealing effects in polymer-nanotube photovoltaic cells. J. Phys. D Appl. Phys. **39**(6), 1058–1063 (2006)

Lau, K.T., Hui, D.: Effectiveness of using carbon nanotubes as nano-reinforcements for advanced composite structures. Carbon **40**, 1605–1606 (2002)

Liang, X., Ling, L., Lu, C., Liu, L.: Resistivity of carbon fibers/ABS resin composites. Mater. Lett. **43**, 144–147 (2000)

Liang, C., et al.: One-step growth of silica nanotubes and simultaneous filling with indium sulphide nanorods. J. Mater. Chem. **14**, 248–252 (2004)

Li, Q., Kinloch, I.A., Windle, A.H.: Discrete dispersion of single-walled carbon nanotubes. Chem. Commun. **26**, 3283–3285 (2005)

Liu, Y.J., Chen, X.L.: Continuum models of carbon nanotube-based composites using the boundary element method. Electron. J. Bound. Elem. **1**, 316–335 (2003a)

Liu, Y.J., Chen, X.L.: Evaluations of the effective material properties of carbon nanotube-based composites using a nanoscale representative volume element. J Mech. Mater. **35**, 69–81 (2003b)

Liu, Y., Nishimura, N., Otani, Y.: Large-scale modeling of carbon-nanotube composites by a fast multipole boundary element method. Comp. Mater. Sci. **34**, 173–187 (2005)

Liu, Z.Y., Xiao, B.L., Wang, W.G., Ma, Z.Y.: Tensile strength and electrical conductivity of carbon nanotube reinforced aluminum matrix composites fabricated by powder metallurgy combined with friction stir processing. J. Mater. Sci. Technol. **30**(7), 649–655 (2014)

Makvandi, R., Öchsne, A.: A refined analysis of the influence of the carbon nanotube distribution on the macroscopic stiffness of composites. Comp. Mater. Sci. **77**(5), 189–193 (2013)

Makvandi, R., Öchsner, A.: On a numerical strategy to simulate nanotube-reinforced composite materials. Mat.-wiss. u. Werkstofftech **45**(5), 429–35 (2014)

Mavalizadeh, S.M., Rahmandoust, M., Öchsner, A.: Numerical investigation of the overall stiffness of carbon nanotubebased composite materials. J. Nano Res.-SW **13**, 47–59 (2011)

McCarthy, B., et al.: A microscopic and spectroscopic study of interactions between carbon nanotubes and a conjugated polymer. J. Phys. Chem. B **106**(9), 2210–2216 (2002)

Moisala, A., Li, Q., Kinloch, I.A., Windle, A.H.: Thermal and electrical conductivity of single- and multi-walled carbon nanotube-epoxy composites. Compos. Sci. Technol. **66**, 1285–1288 (2006)

Mora, R.J., Vilatela, J.J., Windle, A.H.: Properties of composites of carbon nanotube fibers. Compos. Sci. Technol. **69**, 1558–1563 (2009)

Mori, T., Tanaka, K.: Average stress in the matrix and average elastic energy of materials with misfitting inclusions. Acta Metall. **21**, 571–574 (1973)

Mura, T.: Micromechanics of defects in solids, 2nd edn. Martinus Nijhoff Publisher, Leiden (1987)

Nan, C.W., Shi, Z., Lin, Y.: A simple model for thermal conductivity of carbon nanotube-based composites. Chem. Phys. Lett. **375**, 666–669 (2003)

Nobel-Prize: The Nobel Prize in Physics. http://www.nobelprize.org/nobel_prizes/physics/laureates/2010/ (2010)

O'Connell, M.J., et al.: Reversible water-solubilization of single-walled carbon nanotubes by polymer wrapping. Chem. Phys. Lett. **342**(3), 265–271 (2001)

Odegard, G.M., Gates, T.S., Nicholson, L.M., Wise, K.E.: Equivalent continuum modeling of nano-structured materials. NASA TM. **62**, 1869–1880 (2001)

Pipes, R.B., Hubert, P.: Helical carbon nanotube arrays: mechanical properties. Compos. Sci. Technol. **62**, 419–428 (2002)

Ramamurthy, P.C., et al.: Polyaniline/single-walled carbon nanotube composite electronic devices. Solid State Electron. **48**(10), 2019–2024 (2004)

Ren, Y., Li, F., Cheng, H.-M., Liao, K.: Tension-tension fatigue behavior of unidirectional single-walled carbon nanotube reinforced-epoxy composite. Carbon **41**, 2159–2179 (2003)

Ruan, S.L., Gao, P., Yang, X.G., Yu, T.X.: Toughening high performance ultrahigh molecular weight polyethylene using multiwalled carbon nanotubes. Polymer **44**(19), 5643–5654 (2003)

Rul, S., et al.: Percolation of single-walled carbon nanotubes in ceramic matrix nanocomposites. Acta Mater. **52**, 1061–1067 (2004)

Samal, S.S., Bal, S.: Carbon nanotube reinforced ceramic matrix composites—a review. J. Miner. Mater. Charact. Eng. **7**(4), 355–370 (2008)

Sandler, J.K.W., et al.: Ultra-low electrical percolation threshold in carbon-nanotube-epoxy composites. Polymer **44**, 5893–5899 (2003)

Seo, M.-K., Park, S.-J.: Electrical resistivity and rheologycal behavior of carbon nanotubes-filled polypropylene composites. Chem. Phys. Lett. **395**, 44–48 (2004)

Shokrieh, M.M., Rafiee, R.: Investigation of nanotube length effect on the reinforcement efficiency in carbon nanotube based compositesInvestigation of nanotube length effect on the reinforcement efficiency in carbon nanotube based composites. Compos. Struct. **92**, 2415–2420 (2010)

Sie, C.T.: Carbon nanotube reinforced composites: metal and ceramic matrices. Wiley, Weinheim (2009)

Singh, I.V., Tanaka, M., Endo, M.: Effect of interface on the thermal conductivity of carbon nanotube composites. Int. J. Therm. Sci. **46**, 842–847 (2007)

Sinnott, S.B., Ni, B., Mikulski, P.T., Harrison, J.A.: Compression of filled carbon nanotubes: predictions from molecular dynamics simulations. In: Ninth Foresight Conference on Molecular Nanotechnology Proceeding, (2004)

Song, Y.S., Youn, J.R.: Evaluation of effective thermal conductivity for carbon nanotube/polymer composites using control volume finite element method. Carbon **44**, 710–717 (2006)

Spanggaard, H., Krebs, F.C.: A brief history of the development of organic and polymeric photovoltaics. Sol. Energ. Mat. Sol. C. **83**, 125–146 (2004)

Sridhar, I., Narayana, K.R.: Processing and characterizations of MWCNT reinforced aluminum matrix composites. J. Mater. Sci. **44**, 1750–1756 (2009)

Star, A., Lu, Y., Bradley, K., Gruner, G.: Nanotube optoelectronic memory devices. Nano Lett. **4**(9), 1587–1591 (2004)

Tandon, G.P., Weng, G.J.: Average stress in the matrix and effective moduli of randomly oriented composites. Compos. Sci. Technol. **72**(2), 111–132 (1986)

Tang, W., Santare, M.H., Advani, S.G.: Melt processing and mechanical property characterization of multi-walled carbon nanotube/high density polyethylene (MWNT/HDPE) composite films. Carbon **41**, 2779–2785 (2003)

Tucker, C.L., Liang, E.: Stiffness predictions for unidirectional shortfiber composites: review and evaluation. Compos. Sci. Technol. **59**, 655–671 (1999)

Uddin, S.M., et al.: Effect of size and shape of metal particles to improve hardness and electrical properties of carbon nanotube reinforced copper and copper alloy composites. Compos. Sci. Tehnol. **70**, 2253–2257 (2010)

University-of-York: Essential Chemical Industry. http://www.essentialchemicalindustry.org/materials-and-applications/composites.html (2013)

Wagner, H., Lourie, O., Feldman, Y., Tenne, R.: Stress-induced fragmentation of multiwall carbon nanotubes in a polymer matrix. Appl. Phys. Lett. **72**, 188–190 (1998)

Wang, C.M., Zhang, Y.Y., Xiang, Y., Reddy, J.N.: Recent studies on buckling of carbon nanotubes. Appl. Mech. Rev. **63**, 030804-1–030804-18 (2010)

Wei, C.: Thermal expansion and diffusion coefficients of carbon nanotube-polymer composites. Nano Lett. **2**(6), 647–650 (2002)

Weisenberger, M.C., et al.: Enhanced mechanical properties of polyacrylonitrile/multiwall carbon nanotube composite fibers. J. Nanosci. Nanotechnol. **3**(6), 535–539 (2003)

Xie, S., et al.: Mechanical and physical properties of carbon nanotube. J. Phys. Chem. Solids **61**, 1153–1158 (2000)

Xie, X.L., Mai, Y.W., Zhou, X.P.: Dispersion and alignment of carbon nanotubes in polymer matrix: a review. Mat. Sci. Eng. R. **49**, 89–112 (2005)

Zhong, R., Cong, H., Hou, P.: Fabrication of nano-Al based composites reinforced by single walled carbon nanotubes. Carbon **41**, 848–851 (2002)

Zhu, L., Nart, K.A.: Numerical simulation of the effect of nanotube orientation on tensile modulus of carbon-nanotube-reinforced polymer composites. Polym. Int. **53**, 1461–1466 (2004)

Zoo, Y.S., An, J.W., Lim, D.P., Lim, D.S.: Effect of carbon nanotubes addition on tribological behavior of UHMWPE. Tribol. Lett. **16**, 305–309 (2004)

Chapter 4
CNT-Based Nanocomposites

4.1 Introduction

As one of the major classes of polymeric materials, various types of epoxies are used extensively in different engineering applications such as automotive and electronic industries. Epoxy-based composite materials have become proper substitutes for traditional materials like metals, metal alloys, wood, etc. due to their prominent properties such as lightness, ease of processing and relatively low cost. However, one of the major drawbacks in their increasing applications is their poor surface properties.

In the last decade, significant steps have been taken in the field of nanotechnology resulting in new materials with astonishing characteristics. The nanomaterials provided the opportunity to access enhanced macroscopic engineering properties that were seldom available before. Among the discovered nanofillers, carbon nanotubes attracted researchers owing to their extraordinary mechanical, electrical, and thermal properties. Unique atomic structure, very high aspect ratio and extraordinary mechanical properties make carbon nanotubes ideal reinforcing materials in nanocomposites.

Several studies have been accomplished on the effect of CNTs on various properties of polymers. It has been shown that adding a small amount of multi-walled carbon nanotubes would sometimes enhance noticeably the polymer properties such as mechanical and electrical behaviors. The properties of the CNT reinforced nanocomposites depend on a variety of parameters including CNT geometry and the interphase between the matrix and CNT. Also the dispersion of nanomaterials is a combination of chemical and mechanical effects. While it is desirable to have a certain attraction between individual nanotubes to achieve percolation and other property enhancements, it is not desirable that this interaction be so strong as to cause clumping or non-dispersing behavior and thus defeat the very purpose of using nanotubes.

The original version of this chapter was revised. The Erratum to this chapter is available at 10.1007/978-3-319-00251-4_7

M. Rahmandoust and M.R. Ayatollahi, *Characterization of Carbon Nanotube-Based Composites Under Consideration of Defects*, Advanced Structured Materials 39,
DOI 10.1007/978-3-319-00251-4_4

In this chapter processing methods and mechanical, fracture, tribological and electrical properties of CNT-based nanocomposites are studied. Several parameters that affect properties of nanocomposites are discussed. However, defects of preparation of CNT-based nanocomposites and properties enhancement are studied.

4.2 Production Methods

The uniform dispersion of nanoparticles within the nanocomposite and the aggregate prevention are among major concerns in the production process of nanocomposites.

It is known well that the uniform dispersion of nanomaterials is a combination of chemical and mechanical effects. While it is appealing to have a certain attraction between individual nanotubes to attain percolation and other property enhancements, it is not desirable that this interaction be so strong as to cause clumping or non-dispersing behavior.

4.2.1 Dispersion Methods

Common methods to disperse the nanotubes in the matrix are follows:

- Sonication

Ultrasonic devices have locally a high impact of energy, but introduce low portion of shear forces, hence this method is only suitable for very low viscous matrix materials and small volumes. Also, the local introduction of the energy leads to a rupture and damage of CNTs reducing the overall aspect ratio. A proper way to apply the sonication technique to produce CNT-nanocomposites is to disperse them in an appropriate solvent (i.e., acetone), which first allows the agglomerates to be separated due to vibrational energy. The suspension can later be mixed with the epoxy; while simple heating removes the solvent via evaporation.

Fiedler et al. used sonication method to disperse CNT in CNT/epoxy composites. MWCNTs were sonicated for 20 min at 30 % amplitude, mixed with the epoxy resin and again sonicated for 10 min. After that agglomerates were still remaining (Fig. 4.1a). For nanotubes exhibiting a much larger specific surface area (SSA) like DWCNTs the agglomerates are much bigger (Fig. 4.1b). During this sonication process, the samples need to be cooled to avoid an untimely curing of the composite (Fiedler et al. 2006). A reduction of the agglomerate size can be achieved with a functionalisation process, i.e. a combination of sonication and an oxidative process (Gojny et al. 2003). The functional groups, which develop on the surfaces on the CNTs, lead to steric hindering and electrostatic interactions with the solvent, leading to a better distribution of the CNTs in the epoxy matrix as shown in Fig. 4.1b (Fiedler et al. 2006).

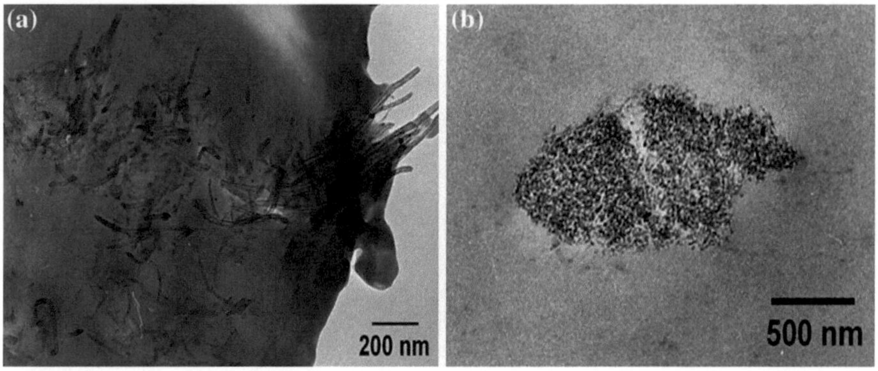

Fig. 4.1 Sample TEM pictures of CNT/epoxy composites: **a** sonication of MWCNT/epoxy; **b** sonication of DWCNT/epoxy (Fiedler et al. 2006)

- Stirrer

This method can be considered as a preliminary method of distributive mixing in a polymer system. This method is good for evaluation on a laboratory scale but the question of scalability to larger processing equipment has to be considered. The size and shape of the propeller and the stirring speed can control distributed quality. In this method, The MWCNTs can be distributed more easily as DW- and SWCNTs, however, they have a tendency to re-agglomerate. This experimentally observed flocculation behaviour is mainly caused by frictional contacts and elastic interlocking mechanisms (Schmid and Klingenberg 2000). Other parameters, like sliding forces and weak attractive forces have only little effect on the flocculation during stirring (Switzer and Klingenberg 2004). In case of the much thinner DW- and SWCNTs, higher shear forces are needed to achieve a fine dispersion.

- Calendaring

The application of a mini-calender to disperse carbon nanotubes and nano-particles in general became a very promising approach to reach a good state of dispersion (Gojny et al. 2004). This method is an established and common technique to disperse micro-particles in different matrices. A major advantage of this method is, besides the improved dispersion results, the efficient manufacturing of larger amounts of nano-composites.

Work by Fiedler et al. shows that the gap size between the Alumina rolls was 5 mm and speed was set to 20 rpm (first roll), 60 rpm (second roll) and 180 rpm (third roll). The dwell time for each batch of suspension on the rolls is in the range of minutes. A first primary dispersion of the agglomerates is performed in the knead-vortexes, whereas the final distribution of the nanoparticles occurs in the thin gap between the rolls. The suspension was collected, mixed with the hardener for 10 min by intense stirring, cured for 24 h at room temperature (RT) and finally post-cured at 60 °C for 24 h. Adding the hardener and further processing the composite according to the manufacturer regulations the final level of

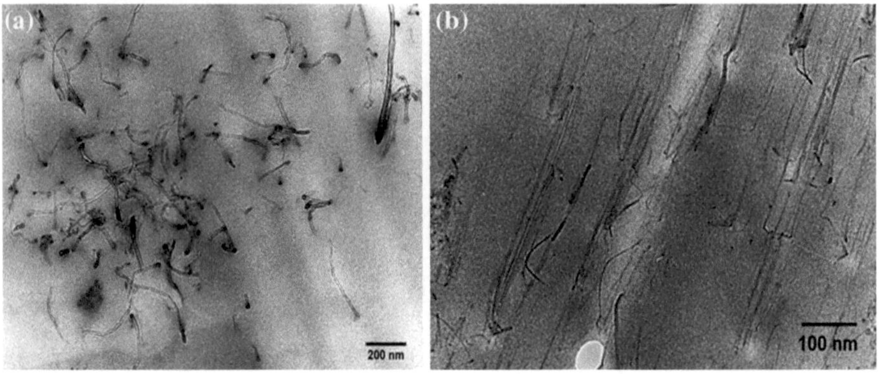

Fig. 4.2 Sample TEM pictures of CNT/epoxy composites: **a** calendering of MWCNT/epoxy; **b** calendering of DWCNT/epoxy (Fiedler et al. 2006)

dispersion was achieved. From Fig. 4.8a, b the high quality of the dispersion can be seen. Figure 4.2 shows that, only a few small aggregates are remaining, which are smeared out and well penetrated by the epoxy matrix (Fiedler et al. 2006).

4.2.2 Composite Processing

The effective utilization of CNT material in composite applications depends strongly on their ability to be dispersed individually and homogeneously within a matrix. An optimized interfacial interaction between the CNT sidewalls and the matrix should result in an efficient load transfer to the "hard" component of the composite. All processing methods known thus far for CNT-based nanocomposites are described below.

4.2.2.1 Solution Processing of Composites

Perhaps the most common method for preparing polymer nanotube composites has been to mix the nanotubes and polymer in a suitable solvent before evaporating the solvent to form a composite film. One of the benefits of this method is that agitation of the nanotubes powder in a solvent facilitates nanotube de-aggregation and dispersion.

Almost all solution processing methods are variations on a general theme which can be summarised as

1. Dispersion of nanotubes in either a solvent or polymer solution by energetic agitation.
2. Mixing of nanotubes and polymer in solution by energetic agitation.
3. Controlled evaporation of solvent leaving a composite film.

In general, agitation is provided by magnetic stirring, shear mixing, reflux or, most commonly, ultrasonication. Sonication can be provided in two forms, mild sonication in a bath or high-power sonication using a tip or horn (Coleman et al. 2006).

4.2.2.2 Melt Processing of Bulk Composites

While solution processing is a valuable technique for both nanotube dispersion and composite formation, it is completely unsuitable for many polymer types that are insoluble. Melt processing is a common alternative, which is particularly useful for dealing with thermoplastic polymers. This range of techniques makes use of the fact that thermoplastic polymers soften when heated. Amorphous polymers can be processed above their glass transition temperature while semi-crystalline polymers need to be heated above their melt temperature to induce sufficient softening. Advantages of this technique are its speed and simplicity, not to mention its compatibility with standard industrial techniques.

In general, melt processing involves the melting of polymer pellets to form a viscous liquid. Any additives, such as carbon nanotubes can be mixed into the melt by shear mixing. Bulk samples can then be fabricated by techniques such as compression molding, injection molding or extrusion. However it is important that processing conditions are optimised, not just for different nanotube types, but for the whole range of polymer–nanotube combinations. This is because nanotubes can affect melt properties such as viscosity, resulting in unexpected polymer degradation under conditions of high shear rates (Coleman et al. 2006).

4.2.2.3 Processing of Composites Based on Thermosets

The most common thermosetting polymers used in the formation of polymer nanotube composites have been epoxy resins. Generally these are polymers that cure when mixed with a catalyzing agent or hardener. In most cases the epoxy begins life in liquid form, facilitating nanotube dispersion by the techniques described in Sect. 4.2.1. Curing is then carried out to convert the liquid composite to the final solid state (Coleman et al. 2006).

In the simplest cases nanotubes have been dispersed by sonication in a liquid epoxy such as Shell EPON 828 epoxy (Ajayan et al. 2000). This blend is then cured by the addition of hardeners such as triethylenetetramine as reported by Ajayan et al. (2000). After having been left to gel overnight, the composite is cured at 100 °C for 2 h.

4.2.2.4 In Situ Polymerization

Since 1999, in situ polymerization of vinyl monomers in the presence of CNT material has been extensively studied for the preparation of functional composites.

The main advantage of this method is that it produces polymer-grafted tubes, mixed with free polymer chains. Moreover, due to the small size of monomeric molecules, the homogeneity of the resulting composite adducts is much higher than mixing CNTs and polymer chains in solution. In this sense, the method allows the preparation of composites with high CNT weight fraction (Spitalsky et al. 2010).

Initially, in situ radical polymerization was applied for the synthesis of PMMA composites by Jia et al. (1999). In this work, the reaction was performed using radical initiator 2,2′-azobisisobutyronitrile (AIBN). The authors suggested that π-bonds of the CNT graphitic network were opened by the radical fragments of initiator and therefore the carbon nanostructures could participate in PMMA polymerization by acting as efficient radical scavengers. If all reactants were mixed simultaneously, the growth of polymer chains was inhibited since many of the initiator molecules would be consumed by the CNTs. By using an improved in situ process (Jia et al. 1999), in which CNT material is added a little time after the mixing of initiator and monomer, longer polymer chains were obtained in the resulting composite, giving rise to better mechanical properties of the composites.

4.2.3 Defects in Nano-composites and Prevention Methods

The challenge in incorporating carbon nanotubes into polymers arises from several perspectives. Because of their high aspect ratio, carbon nanotubes are capable of entanglements and as a consequence they will be expected to give very high increases in the system viscosity as the individual nanotubes un-bundle, one from the other. A second consideration is that the individual tubes are difficult to separate from each other as a result of Van der Waals forces that tend to make them clump together.

Work by Hough et al. shows that at a volume fraction of 0.0026, aqueous suspensions of single-walled nanotubes (SWCNTs) exhibit the onset of solid-like elasticity. Development of a master curve for shear-dependent stresses enables derivation of an inter-nanotube interaction energy of 40 k_BT per bond. One may postulate that SWCNTs form interconnected networks in suspension with bonds that freely rotate and resist stretching. The observed elasticity originates from bonds between the nanotubes rather than from stiffness or stretching of the individual tubes (Hough et al. 2004).

The stress is mechanically transferred from the surrounding polymer matrix to the CNT through shear stress build-up at interface between the CNT and surrounding polymer matrix. Numerous researchers have attributed lower-than-predicted CNT-polymer composite properties to a lack of interfacial bonding (Weisenberger et al. 2003).

The surface of a CNT is essentially an exposed graphene sheet. The weak interplanar interaction of graphite provides its solid lubricant quality, but this in turn can lead to a weakening of the polymer-NT interaction and resistance to matrix adhesion. In the case of MWCNTs, this could be expected to lead to weakening of the polymer-NT interaction. Furthermore, the relatively chemically inert nature of graphene structures does not favor a strong polymer-NT interaction in the case of both MWCNTs and SWCNTs (Cadek 2002).

The high aspect ratio of CNTs is generally a barrier to facile incorporation of nanotubes into polymer media. The high surface energies of the nanotubes render separation into individual tubes difficult but also increase the difficulty of maintaining separated tubes in a polymer media. Several workers have reported methods to chemically functionalize the surface of nanotubes but these methods are generally on a very small scale and have not been scaled up to any scale suitable for commercial activity (Ashton 2010).

Graphene structures have a relatively low chemical reactivity (Aihara 1994). This can lead to a lower than expected interaction between a nanotube and a polymer matrix in which it is embedded. One way to enhance and strengthen this interaction is to chemically change the nature of the nanotube surface.

There have been several published accounts of chemical modification of the surface of carbon nanotubes. These often involve chemical modification of the surface under vigorous or forcing conditions (Chen et al. 1998; Mikelson et al. 1998; Bahr et al. 2001; Pekker et al. 2001; Lin et al. 2003; Sun et al. 2002). Functional groups such as fluorine and hydrogen can be generated that must be removed, and this can often require the use of aggressive chemicals such as alkyl-lithiums in the case of fluorine. Carboxyl groups generated in an oxidizing medium can also be used as sites for further chemistry, such as amidation or transesterification type reactions. These reactions can assist in the solubilization of nanotubes (Sun et al. 2001). Sulfonate groups can be attached to multiwalled nanotubes (MWCNTs) (Dai and Mau 2001). These functionalized materials were compounded into polyaniline to form composites with enhanced electrical conductivity arising from the doping of the polymer by the sulfonated MWCNTs. Another approach to solubilization of nanotubes in water is achieved by wrapping the nanotubes with polyvinylpyrollidone (PVP) or polytheylene imine (PEI) (Shim et al. 2001; Chen et al. 1998).

Incorporation of nanotubes (NTs) into polymer media is rendered difficult by

- High aspect ratio
- Small size
- Tendency to agglomerate or clump.

However, when CNTs can be dispersed effectively, significant toughening of polymer matrices, such as ultrahigh molecular weight polyethylene (UHMWPE), HDPE, PAN, PVA, and iPP, has been reported at relatively low loading levels of 1–5 % (Dalton et al. 2003). Modifications of morphology and also measurements of the energy required for debonding of CNTs suggest that the CNTs may increase the energy absorption or toughness of the composite and lead to prolonged fatigue-life (Cadek 2002).

This would clearly indicate that the transfer of mechanical energy at the nanotube-polymer interface is critical. The mechanisms that may be at work are prolongation of the formation of crazing or microcracking as a result of the nanoparticle acting as a stress concentrator. This mechanism is well known in fracture mechanics. Another possible mechanism is simply physical bridging of a developing crack by a nanotube to prevent further crack propagation. However, further work is needed to fully understand the reinforcing mechanisms at work in carbon nanotube-polymer composites. The nature of these materials indicates that sophisticated analytical techniques will have to be used to understand the mechanisms at work (Ren et al. 2003).

Percolation is a statistical concept concerned with the formation of long-range clusters of connected particles or pathways in random systems. One definition of percolation relates to the slow flow of fluids through a porous medium, but more theoretically it refers to lattice models of random systems and their internal connectivity. In such systems, there is a critical concentration at which the connectivity appears.

This conceptual model applies well to dispersions of nanotubes in polymers. As the concentration of nanotubes progressively increases, there is a concentration where there is a jump in conductivity. This is correlated with the onset of connectivity and this concentration is usually defined as the percolation threshold. Zou et al. showed that for the dispersion of MWCNTs in a polymer matrix by a twin-screw extruder, there is a critical MWCNT concentration of 1.0 wt% where a fine network of filler is formed above which the composite possesses improved mechanical properties (Zou et al. 2004).

Pötschke report a study in which a polycarbonate (PC) masterbatch was progressively diluted by PC, in a conical twin-screw extruder, to afford different nanotube concentrations. Electrical measurements indicated percolation of MWCNT between 1.0 and 1.5 wt% (Potschke et al. 2003).

Nanocomposites comprised of double-wall carbon nanotubes (DWCNTs) and an epoxy matrix were produced by a standard calendering technique. A very good dispersion of both the DWCNT and carbon black (CB) in an epoxy resin was obtained. Fracture-mechanical properties resulted in an increase of strength, Young's modulus, and strain to failure at a nanotube content of only 0.1 wt% (Gojny et al. 2004).

Lowering the percolation threshold is a critical consideration in achieving low loading levels (lower cost) of a conducting medium in an insulator (conducting composites). This has relevance in applications where charge dissipation and electrical conductivity is desired. Rod-shaped particles tend to have lower percolation thresholds rather than lower aspect-ratio particles. Altering the interaction potential between particles also has a significant impact. In the case of nanotubes, there are different measures of percolation thresholds, ranging from concentrations below 1 % to concentrations in the 4–5 % range and above. One interpretation of these results is that the use of dispersants and surfactants can have a significant impact (Gojny et al. 2004).

The correct balance of attraction between individual particles combined with dispersion and distribution in the polymer medium is what is required. In practice, there is large variability of experimental results in the literature dealing with the electrical behavior of composites loaded with conducting rods. Control of the percolation threshold is difficult as numerous factors have an impact (Vigolo et al. 2005).

4.3 Mechanical Properties

The one-dimensional structure of CNTs, their low density, their high aspect ratio, and extraordinary mechanical properties make them particularly attractive as reinforcements in composite materials. By now, a wide range of publications have reported certain aspects of the mechanical enhancement of different polymer systems by CNTs. These studies have been discussed in some excellent reviews. The variation of many parameters, such as CNT type, growth method, chemical pretreatment as well as polymer type and processing strategy has given some encouraging results in fabricating relatively strong CNT–polymer composites (Spitalsky et al. 2010).

With a tensile modulus on the order of 1 TPa and a tensile strength 100 times greater than steel at one sixth the weight, SWCNTs are ideal reinforcements for polymer matrix composites, but to fully exploit their exceptional properties, nanotubes should be homogeneously dispersed and aligned within a polymer matrix. Two main approaches to combining the nanotubes and the polymers are in situ polymerization and compounding of the nanotubes and polymer. In situ polymerization has been shown to produce direct bonding between the nanotubes and the polymer (Jia et al. 1999), and methods of polymerization involving sonication have been shown to produce high dispersion (Park et al. 2002).

Static and dynamic mechanical property improvements have been observed in nanotube-reinforced polymers compared with the neat polymers, and anisotropic nanotube orientation has shown greater improvements over isotropic nanotube orientations in polymers. For example enhancements of ~450 % in tensile modulus have been observed at nanotube loadings of 20 wt% even though functionalization was not used (Haggenmueller et al. 2003).

Ayatollahi et al. studied the effect of carbon nanoreinforcements of different shapes including the spherical (ND), cylindrical fibers (CNT and CNF) and platelets (GO) on the mechanical properties of epoxy-based nanocomposites. According to the experimental results, the ND, CNT and CNF particles provided reasonable enhancement in the tensile properties of nano-composites, while the GO platelets did not show any noticeable increase (Alishahi et al. 2013; Ayatollahi et al. 2011a).

4.3.1 Effects of Composite Processing

In the category of solution-based bulk composites, the most obvious conclusion is that matrices filled with CNTs bearing covalently attached polymer chains show enhanced mechanical properties. This is not surprising since polymer grafting should significantly improve both CNT dispersion and stress transfer to the matrix. Regarding the blending of unmodified CNTs with polymers by a solution-phase protocol, PVA shows significant reinforcement after incorporation of CNT material (Coleman et al. 2004; Ryan et al. 2007). This can be explained by the fact that CNT sidewalls nucleate the crystallization of the polymer, thus leading to stiffer composite (Spitalsky et al. 2010).

In situ polymerization has presented a number of advantages over other composite processing methods (Spitalsky et al. 2010). This is because it is easier to get strong interactions between polymer and nanotube during the growth stage rather than mixing the constituents by various means. In some cases, depending on how the reaction is carried out, some of the polymer strands can even be covalently attached to the nanotubes. Needless to add, that these methods are expected to yield large interfacial shear strengths because of the strong polymer/nanotube interactions. Previous researches show that low reinforcement observed for the melt processed bulk composites (Spitalsky et al. 2010). On the contrary, microfibers produced by melt spinning/drawing techniques display better reinforcement compared to the bulk, due to CNT alignment effects (Dondero and Gorga 2006; Jose et al. 2007).

4.3.1.1 Mechanical Properties of Solution Processed Composites Based on Thermoplastic Polymers

In this section mechanical properties of solution based composites will be presented. Many studies have been published each with a different focus. However, the common theme seems to have been enhancement of Young's modulus. All results pertaining to Young's Modulus are summarised in Table 4.1. In the table Y_{poly} is tensile modulus of neat polymer and Y_{max} is the highest tensile modulus that is achieved by adding carbon nanotube.

Table 4.1 Modulus of solution-based matrices and their composites

Nanotube type	Polymer	Y_{poly} (GPa)	Y_{Max} (GPa)	NT content	References
MWCNT	PVA	~6.3	~12.6	60 wt%	Shaffer and Windle (1999)
MWCNT	PS	1.2	1.69	1 wt%	Qian et al. (2000)
MWCNT	PVA	7	12.6	0.6 vol%	Cadek et al. (2002)
MWCNT	PVK	2	5.6	4.8 vol%	Cadek et al. (2002)
MWCNT	PS	1.53	3.4	2.5 vol%	Safadi et al. (2002)
MWCNT	HDPE	0.98	1.35	1 wt%	Ruan et al. (2003)
MWCNT	MEMA	0.71	2.34	1 wt%	Velasco-Santos et al. (2003a, b)
DWNT	PVA	2	3.6	0.2 vol%	Cadek et al. (2004)
MWCNT	PVA	2	4.2	0.6 vol%	Cadek et al. (2004)
SWCNT	PVA	2	3.2	2 vol%	Cadek et al. (2004)
MWCNT	PVA	1.9	7.04	0.6 vol%	Coleman et al. (2004)
DWNT	PVA	1.28	1.43	0.4 vol%	Khan and Coleman (2007)
MWCNT	PU	1.37	6.02	47.9 vol%	Han et al. (2014)

Table 4.2 Modulus of melt-processed bulk matrices and their composites

Nanotube type	Polymer	Y_{poly} (GPa)	Y_{Max} (GPa)	NT content	References
MWCNT	PMMA	0.73	1.63	17 wt%	Jin et al. (2001)
MWCNT	PS	2	4.5	25 vol%	Andrews et al. (2002a, b)
MWCNT	PS	2.3	2.6	5 wt%	Thostenson and Chou (2002)
MWCNT	PC	0.8	1.04	15 wt%	Potschke et al. (2003)
MWCNT	PMMA	2.7	3.7	10 wt%	Gorga and Cohen (2004)
MWCNT	PA6	2.6	4.2	12.5 wt%	Meincke et al. (2004)
MWCNT	PA6/ABS	1.97	2.51	7.5 wt%	Meincke et al. (2004)
MWCNT	Nylon	0.4	1.24	2 wt%	Liu et al. (2004)
SWCNT	PP	0.85	1.19	0.75 wt%	López Manchado et al. (2005)
SWCNT	pp	1.60	1.99	5 wt%	Bhuiyan et al. (2011)
MWCNT	PP	1.38	1.87	5 wt%	Bhuiyan et al. (2013)

4.3.1.2 Mechanical Properties of Melt Processed Composites

In this section we discuss mechanical properties of bulk composites fabricated by melt processing. All the results quoted for Young's modulus are summarised in Table 4.2.

4.3.1.3 Mechanical Properties of Composites Based on Thermosetting Polymers

Due to their wide range of industrial uses, thermosetting polymers in general and epoxy resins in particular have been widely studied as a potential matrix for nanotube based composites. In this section all matrices were epoxy resins unless otherwise stated. Details of the matrix and a summary of the modulus values are given in Table 4.3.

4.3.2 Effects of CNT and Matrix Type and Filler Loading

A number of studies show that after a critical CNT content in matrix, mechanical properties decrease with increasing critical loading of CNTs and sometimes, these properties decrease below the neat matrix material (Mittal et al. 2015). Chen et al. produced polycarbonate/MWCNT nanocomposites by large scale extruder and studied properties at various concentrations of CNTs into matrix. They observed that a composite containing 1 wt% CNT shows 4.5 % higher in tensile strength than in neat polycarbonate. With increase of MWCNTs content there is a significant decrease in tensile strength or elongation at break. In addition, decrease in

Table 4.3 Modulus of thermoset polymer matrices and that of their respective composites

Nanotube type	Resin/hardener	Y_{poly} (GPa)	Y_{Max} (GPa)	NT content (wt%)	References
MWCNT	Shell EPON 828/ triethylenetetraamine	3.1	3.71	5	Cooper et al. (2001)
MWCNT	EPON SU-8	4.2	5.0	0.1	Xu et al. (2002)
MWCNT	Bisphenol A-epichlorhydrine/ triethylenetetramine	0.118	0.465	4	Allaoui et al. (2002)
MWCNT	Bisphenol A-epichlorhydrine/ triethylenetetramine	1.2	2.4	1	Bai (2003)
MWCNT	D.E.R. 736 P/D.E.H. 24, Dow Chemical	0.0014	0.0021	0.25	(Li et al. 2004a, b)
MWCNT	Polyimide triple A PI	2.84	3.9	14.3	Ogasawara et al. (2004)
MWCNT	LY 5052 CIBA-GEIGY/HY5052CIBA-GEIGY	2.75	4.13	6	Breton et al. (2004)
SWCNT	1,1-Bis(4-cyanatophenyl) ethane	4	7	5	Li et al. (2004a, b)
MWCNT	Epon 862/Epikure W	2.86	3.48	0.5	Chen et al. (2014)
MWCNT	EPON 828/Epilox	0.406	0.723	0.3	S-Hadavand et al. (2013)

fluidity with increasing CNT's loading became an obstacle in uniform microstructure and resulted into lowered mechanical properties (Chen et al. 2007).

Multiscale modeling for mechanical properties of carbon nanotube reinforced nanocomposites subjected to different types of loading was performed by Ayatollahi et al. (2011b). They showed that the stiffness of the nanocomposite could be affected significantly by the properties of interphase between MWCNT and the matrix. In addition, the stiffness of the interphase had the maximum effect on the stiffness of the nanocomposite in the bending loading conditions (Ayatollahi et al. 2011b). This result shows the importance of enhancing the interaction between a nanotube and a polymer matrix to achieve better mechanical properties. This can be achieved for example by chemical modification of the surface of carbon nanotubes.

Ayatollahi et al. studied the effects of multi-walled carbon nanotubes (MWCNTs) on the mechanical properties of epoxy/MWCNT nano-composites. They showed that the mechanical properties of polymers are generally supposed to be improved by addition of CNTs. The results of the tensile tests in this study are presented in Table 4.4 and typical stress-strain curves of the pure epoxy and the nano-composites are shown in Fig. 4.3. In general, higher filler loading resulted in higher Young's modulus of nano-composite (Ayatollahi et al. 2011c).

Table 4.4 Tensile properties of pure epoxy and nano-composites (Ayatollahi et al. 2011c)

Samples	Young modulus E_C (GPa)	Tensile strength σ_c (MPa)	Elongation at break ε_t (%)
Pure epoxy	3.15 ± 0.14	68.35 ± 2.43	3.10 ± 0.17
0.1 wt% MWCNT	3.24 ± 0.25	71.74 ± 3.32	2.98 ± 0.3
0.5 wt% MWCNT	3.51 ± 0.18	73.94 ± 6.17	2.87 ± 0.27
1 wt% MWCNT	3.69 ± 0.30	72.46 ± 5.24	2.70 ± 0.31

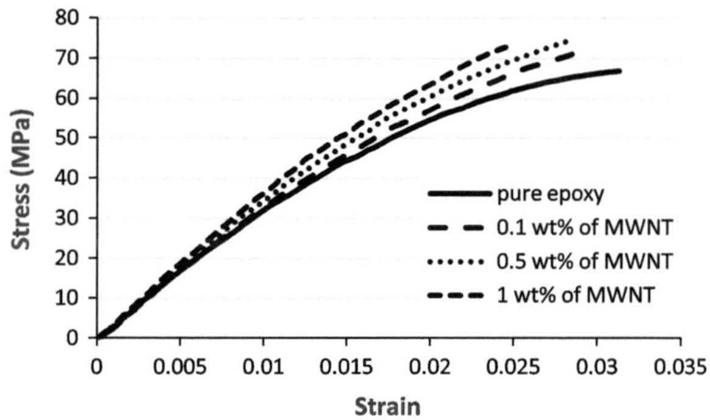

Fig. 4.3 Typical stress-strain curves for CNT-based polymer composites with different MWCNT weight percents (Ayatollahi et al. 2011c)

The tensile strength values were also improved by adding MWCNTs to epoxy. However, the improvement in tensile strength occurred only when MWCNT was added up to 0.5 wt%. The presence of more agglomerates in the specimen with 1 wt% of MWCNTs, could be the reason for this trend (Ayatollahi et al. 2011c).

Table 4.4 also shows a reduction in the elongation at break of nano-composites relative to that of pure epoxy, indicating a ductile to brittle trend in nano-composite with increasing the filler loading (Ayatollahi et al. 2011c). A similar finding has been reported by Satapathy et al. (2007) as well. It is noteworthy that different flaws like voids and impurities can reduce the elongation at break as well.

Table 4.5 represents an extensive collection of published data in this field. The table is organized as follows:

In column 1, CNT species (single/double/multi-walled) and state (pristine/oxidized/chemically modified) are given. The polymer matrices in column 2 are arranged by type. Finally, in columns 3 the weight fraction of CNTs is specified.

Mechanical characteristics like tensile strength at break, elastic modulus and toughness of the composite materials are listed in columns 4–6, respectively (% change compared to the neat matrix). In addition, the % change of storage modulus of composites is given in column 7 (in cases where DMA technique was used). Finally, the type of mechanical testing (tensile, DMA, etc.) and the reference number are listed in columns 8 and 9, respectively. Inspection of Table 4.4 clearly shows a large spread in the number of investigations for different polymer matrices.

Table 4.5 Mechanical properties of CNT-based polymer composites

CNT type	Matrix	CNT weight fraction (%)	Composite tensile strength (% increase)	Composite Young modulus (% increase)	Toughness (% increase)	Composite storage modulus (% increase)	Mechanical testing	References
Purified MWCNTs	PS	40	10	100	–	100	Tensil, DMA	Andrews et al. (2002a, b)
Pristine MWCNTs	PS	5	–	–	–	40	DMA	Thostenson and Chou (2002)
CPP-grafted MWCNTs	PS	0.5	100	100	–	–	Tensile	Blake et al. (2006)
SCMS-grafted SWCNTs	PS	0.06	82	78	–	–	Tensile	Xie et al. (2007)
Pristine MWCNTs	UHMWPE	15	–	–	–	−50	DMA	Bin et al. (2003)
Purified MWCNTs	UHMWPE	1	25	0	–	–	Tensile	Ruan et al. (2003)
Purified MWCNTs	UHMWPE	1	10	15	–	–	Tensile	Wang et al. (2005)
Oxidized MWCNTs	UHMWPE	5	19	12	44	–	Tensile	Ruan et al. (2006)
Pristine MWCNTs	HDPE	5	12	10	5	–	Punch test	Tang et al. (2003)
Modified MWCNTs	HDPE	1	20	50	–	–	Tensile	Zou et al. (2004)
PE-grafted SWCNTs	HDPE	0.5	−2	0	15	–	Tensile	Tong et al. (2004)
Oxidized MWCNTs	HDPE	0.44	5	22	32	–	Tensile	Kanagaraj et al. (2007)
Fluorinated SWCNTs	MDPE	5	–	–	–	30	DMA	Shofner et al. (2006)
Pristine MWCNTs	MDPE	1	11	19	–	–	Tensile	Kearns and Shambaugh (2002)
Pristine MWCNTs	LDPE	10	56	89	–	–	Tensile	Xiao et al. (2007a, b)
Purified MWCNTs	LDPE	3	150	30	100	–	Tensile	Gorrasi et al. (2007)
PE-grafted MWCNTs	LDPE	0.5	30	65	60	–	Tensile	Yang et al. (2007)

(continued)

Table 4.5 (continued)

CNT type	Matrix	CNT weight fraction (%)	Composite tensile strength (% increase)	Composite Young modulus (% increase)	Toughness (% increase)	Composite storage modulus (% increase)	Mechanical testing	References
Pristine SWCNTs	PP	1	40	55	–	–	Tensile	Kearns and Shambaugh (2002)
Purified SWCNTs	PP	1.8	–	–	–	15	DMA	Grady et al. (2002)
Purified MWCNTs	PP	1	20	15	–	–	Tensile	Xia et al. (2004)
Pristine SWCNTs	PP	1	45	–	–	42	Tensile, DMA	Moore et al. (2004)
Pristine SWCNTs	PP	1	150	200	–	–	Tensile	Chang et al. (2005)
Pristine SWCNTs	PP	0.75	15	40	–	60	Tensile, DMA	Lopez Manchado et al. (2005)
Pristine SWCNTs	PP	0.5	–	–	–	75	DMA	Leelapornpisit et al. (2005)
Pristine MWCNTs	PP	0.25	10	130	30	–	Tensile	Dondero and Gorga (2006)
Fluorinated SWCNTs	PP	10	150	110	–	–	Tensile	McIntosh et al. (2006)
Oxidized MWCNTs	PP	0.6	20	300	–	–	Tensile	Xiao et al. (2007a, b)
Pristine MWCNTs	PP	0.3	40	40	–	–	Tensile	Zhao et al. (2007)
Purified SWCNTs	PP	1	400	270	–	–	Tensile	Jose et al. (2007)
Pristine MWCNTs	PP	10	170	130	–	–	Tensile	McIntosh et al. (2007)
Pristine SWCNTs	PP	0.9	3	60	–	–	Tensile	Masuda and Torkelson (2008)
Purified SWCNTs	PMMA	8	50	100	–	–	Tensile	Haggenmueller et al. (2000)
Pristine MWCNTs	PMMA	26	–	–	–	100	DMA	Jin et al. (2001)

(continued)

Table 4.5 (continued)

CNT type	Matrix	CNT weight fraction (%)	Composite tensile strength (% increase)	Composite Young modulus (% increase)	Toughness (% increase)	Composite storage modulus (% increase)	Mechanical testing	References
PVDF-coated MWCNTs	PMMA	2.2	–	–	–	140	DMA	Jin et al. (2002)
SWCNTs	PMMA	0.1	–	4	–	–	Tensile	Cooper et al. (2002)
MWCNTs	PMMA	4	–	−5	–	–	Tensile	Cooper et al. (2002)
Oxidized MWCNTs	PMMA	1.5	75	–	75	88	Tensile, DMA	Velasco-Santos et al. (2003a, b)
Purified SWCNTs	PMMA	2	–	90	–	–	Tensile	Du et al. (2003)
Purified SWCNTs	PMMA	0.014	–	–	–	10	DMA	Putz et al. (2004)
Pristine MWCNTs	PMMA	1	0	0	170	–	Tensile	Gorga and Cohen (2004)
PMMA-grafted MWCNTs	PMMA	20	–	–	–	1100	DMA	Huang (2004)
Modified SWCNTs	PMMA	1	0	0	100	–	Tensile	Sabba and Thomas (2004)
PMMA-grafted MWCNTs	PMMA	0.15	360	90	1200	–	Tensile	Blond et al. (2006)
Modified MWCNTs	PMMA	15	–	–	–	60	DMA	Wang et al. (2006a, b)
Octyl-modified MWCNTs	PMMA	5	160	260	–	–	AFM	Liu et al. (2007a, b, c)
Octyl-modified MWCNTs	PMMA	2.1	0	0	200	–	Tensile	Liu et al. (2007a, b, c)
Modified MWCNTs	PMMA	0.1	90	100	–	–	Tensile	Kim and Jo (2008)
Oxidized MWCNTs	PVA	60	–	–	–	100	DMA	Shaffer and Windle (1999)

(continued)

Table 4.5 (continued)

CNT type	Matrix	CNT weight fraction (%)	Composite tensile strength (% increase)	Composite Young modulus (% increase)	Toughness (% increase)	Composite storage modulus (% increase)	Mechanical testing	References
PVP-coated SWCNTs	PVA	5	80	100	–	–	Tensile	Zhang et al. (2003)
Purified MWCNTs	PVA	1	330	270	70	–	Tensile	Coleman et al. (2004)
Thin MWCNTs	PVA	1	–	100	–	–	Tensile	Cadek et al. (2004)
Purified SWCNTs	PVA	2.5	−5	35	–	–	Tensile	Paiva et al. (2004)
Purified SWCNTs	PVA	0.1	–	–	–	30	DMA	Probst et al. (2004)
Purified SWCNTs	PVA	3	22	40	–	–	Tensile	Zhang et al. (2004)
Hydroxy-modified SWCNTs	PVA	0.8	47	79	–	–	Tensile	Liu et al. (2005)
Ferritin-grafted MWCNTs	PVA	1.5	–	–	–	105	DMA	Bhattacharyya et al. (2005)
Collagen-modified SWCNTs	PVA	4	300	260	700	–	Tensile	Bhattacharyya et al. (2006)
Pristine SWCNTs	PVA	0.2	100	200	–	–	Tensile	Ryan et al. (2007)
Oxidized MWCNTs	PVA	0.2	52	46	–	–	Tensile	Hou et al. (2009)
Oxidized SWCNTs	PVA	0.2	42	29	–	–	Tensile	(Hou et al. 2009)
Pristine MWCNTs	Epoxy	3	50	100	–	–	Tensil	Park et al. (2003)
Purified SWCNTs	Epoxy	5	7	0	–	–	Tensile	Kim et al. (2004a, b)
Oxidized MWCNTs	Epoxy	6	−25	32	–	–	Tensile	Breton et al. (2004)
Pristine SWCNTs	Epoxy	31.3	–	–	–	492	DMA	Wang et al. (2004)
Pristine DWCNTs	Epoxy	0.1	−3	2	18	–	Tensile	Gojny et al. (2004)
Purified SWCNTs	Epoxy	0.5	–	–	–	51	DMA	Liao et al. (2004)

(continued)

Table 4.5 (continued)

CNT type	Matrix	CNT weight fraction (%)	Composite tensile strength (% increase)	Composite Young modulus (% increase)	Toughness (% increase)	Composite storage modulus (% increase)	Mechanical testing	References
Fluorinated SWCNTs	Epoxy	0.3	–	–	–	20	DMA	Miyagawa and Drzal (2004)
Copolymer-modified MWCNTs	Epoxy	0.25	50	50	–	–	Tensile	Li et al. (2004a, b)
NH2-modified DWCNTs	Epoxy	0.5	10	15	45	–	Tensile	Gojny et al. (2005)
NH2-modified SWCNTs	Epoxy	0.5	–	–	–	25	DMA	Wang et al. (2006a, b)
MA-modified MWCNTs	Epoxy	1	45	100	–	–	Tensile	Tseng et al. (2007)
Oxidized MWCNTs	Epoxy	4	40	−10	–	–	Tensile	Guo et al. (2007)
MWCNTs	Epoxy	0.5	62	54	–	–	Tensile	Spitalsky et al. (2009a, b)
NH2-modified SWCNTs	Epoxy	0.1	12	56	–	–	Tensile	Valentini et al. (2008)
Oxidized MWCNTs	Epoxy	1	–	–	–	75	DMA	Spitalsky et al. (2009a, b)
PEI coated MWCNTs	Epoxy	1	–	–	–	8	DMA	Liu et al. (2009)
NH2-modified MWCNTs	Epoxy	0.6	–	–	–	9	DMA	Yang et al. (2009)
MWCNT	PS	3	–	43	–	–	Tensile	Lahelin et al. (2011)
MWCNT	PMMA	15	–	24	–	–	Tensile	Lahelin et al. (2011)

4.3.3 Effects of CNT Aspect Ratio

Work by Ayatollahi et al. shows that the mechanical properties have great dependency on filler dimensions such that at higher aspect ratios the mechanical properties are generally improved (Ayatollahi et al. 2011d). In order to determine the suitable amount of filler, they prepared a number of test specimens based on different MWCNTs contents ($30 < d < 40$) namely 0.1, 0.5 and 1 wt%. The results of tensile tests are presented in Table 4.6. Table 4.7 indicates different types of MWCNTs used in their study.

As can be seen in Table 4.6, the best Young's modulus was obtained for 1 wt% of MWCNT, while the highest tensile strength was achieved for 0.5 wt%. Therefore, 0.5 wt% of MWCNT was chosen as the filler content for the rest of this study to prevent the effect of aggregates on the mechanical properties. In the next stage, epoxy/0.5 wt% MWCNT nanocomposites of all other three types of MWCNTs were prepared and tested. The results presented in Fig. 4.4 show that with increasing the aspect ratio of MWCNTs, the modulus of elasticity and the tensile strength both increase.

In order to explain the trends observed for the tensile properties, the effective mechanisms which contribute to these properties were studied. The load transfer properties are the main issue that ensures the composite system functions properly. Since the reinforcement role of MWCNTs depends on the load transfer from matrix to MWCNT, the interfacial strength between these two is of crucial importance. It has been already shown that cylindrical particles with smaller diameter build a stronger interface with the matrix (Barber et al. 2004; Manoharan et al. 2009). Stronger interface leads to a better load transfer, which causes improved mechanical properties.

Table 4.6 Mechanical properties of pure epoxy and nano-composites reinforced with MWCNTs of type B (Ayatollahi et al. 2011d)

Samples	Young's modulus E_C (GPa) (Exp.)	Tensile strength σ_c (MPa)	Elongation at break ε_t (%)	Fracture toughness (MPa $m^{0.5}$)
Pure epoxy	3.15 ± 0.14	68.35 ± 2.43	3.10 ± 0.17	1.62 ± 0.14
0.1 wt% MWCNT	3.24 ± 0.25	71.74 ± 3.32	2.98 ± 0.3	1.86 ± 0.07
0.5 wt% MWCNT	3.51 ± 0.18	73.94 ± 6.17	2.87 ± 0.27	2.05 ± 0.1
1 wt% MWCNT	3.69 ± 0.30	72.46 ± 5.24	2.70 ± 0.31	1.93 ± 0.19

Table 4.7 MWCNTs of different aspect ratios used by Ayatollahi et al. (2011d)

Category	Diameter range (nm)	Length range (μm)	Aspect ratio (L/d)
Type A	20–30	10–30	1000
Type B	30–40	10–30	715
Type C	40–50	10–30	555
Type D	50–60	10–30	455

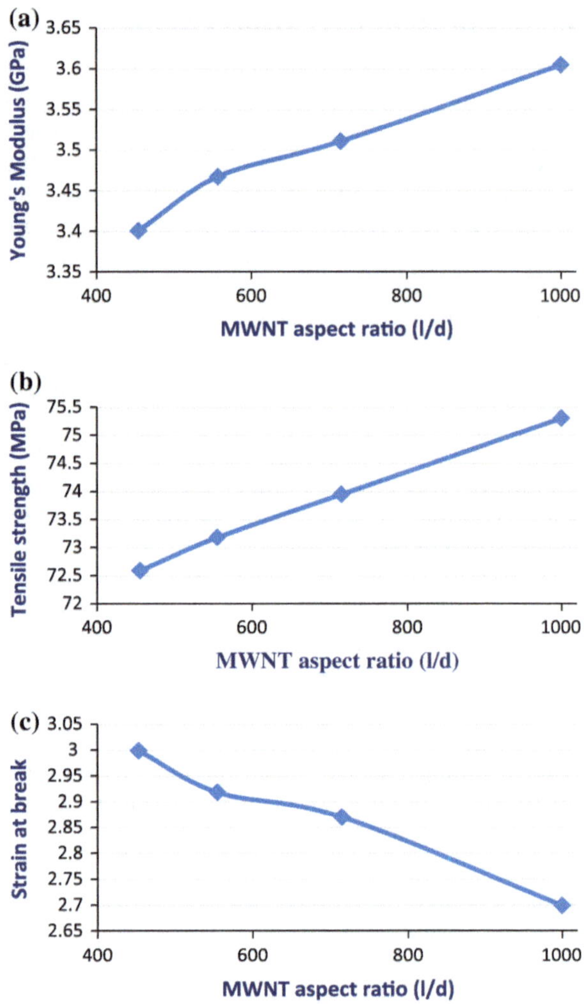

Fig. 4.4 Effects of MWCNT aspect ratio on experimental results obtained for nanocomposites **a** Young's modulus **b** tensile strength **c** strain at break (Ayatollahi et al. 2011d)

In addition, a well-known mechanism which participates in improving the mechanical properties of nanocomposites is bridging. Clearly, for a constant weight fraction of nanotubes, as the tube diameter decreases, the number of nanotubes bridging the crack interface shows an increase (Zhang et al. 2008). More MWCNTs with stronger interface could be a reason for higher tensile strength and fracture toughness in the nanocomposites reinforced with MWCNTs of smaller diameters.

Decrease of strain at break shown in Fig. 4.4c can be attributed to ductile to brittle transition of nanocomposites when the aspect ratio of MWCNTs increases. This is because a higher number of nanotubes can reduce the mobility of polymer chains which leads to more brittle behavior (Ayatollahi et al. 2011d).

4.3.4 Effects of Secondary Nanoparticles

In some studies, the effects of adding synthetic nanoparticles on the mechanical properties of the nanocomposites were investigated. The effect of hybrid particles is mostly positive and can improve the static and dynamic properties of composites. In this category, applying carbon nanotubes (CNTs) with different nanoparticles as hybrid fillers were taken into account in the literature to improve the fatigue behavior, mechanical and electrical properties of reinforced composites.

Boger et al. appointed silica and MWCNT hybrid nanoparticles to increase the high cycle fatigue life of epoxy laminates and finally reported that the life was increased by several orders of magnitude in number of load cycles (Böger et al. 2010). Fritzsche et al. investigated the CNT based elastomer-hybrid-nanocomposites prepared by melt mixing and showed promising results in electrical, mechanical and fracture-mechanical properties (Fritzsche et al. 2009). Witt et al. improved mechanical properties such as tensile strength and strain to failure of a conductive silicone rubber composite using both CNTs and carbon black (CB) (Witt et al. 2013). Al-Saleh Mohammed and WalaaSaadeh fabricated a nanostructured hybrid polymeric materials based on CNTs, CB and CNFs and investigated electrical properties and electromagnetic interference shielding effectiveness in the X-band frequency range (Al-Saleh and Saadeh 2013). Jen et al. applied hybrid Magnesium/carbon fiber to increase the fatigue life of nanocomposite laminates (Jen 2012).

Rahmanian et al. synthesized multiscale fillers through growing carbon nanotubes (CNT) on silica microparticles via catalytic chemical vapour deposition process. The CNT-silica fillers were incorporated within polypropylene (PP) matrix by melt-mixing process. An improvement by more than 35 % was achieved for elastic modulus and tensile strength of nanocomposites (Rahmanian et al. 2014).

Applying carbon nanotubes (CNT) and graphite nanoplatelets (GNPs) to epoxy nanocomposites was shown by Li et al. (2008). In this research the effects of hybrid carbon nanotubes (CNT) and graphite nanoplatelets (GNPs) on the mechanical and electrical properties of epoxy nanocomposites were studied. It was represented that While the flexural properties do not change much due to the hybridization, the fracture toughness increases by 21 % compared to the nanocomposite with 2 wt% GNP alone, or 57 % compared to neat epoxy. Crack tip bridging and pull out of CNTs are identified as the main toughening mechanisms of hybrid nanocomposites, in addition to the existing crack tip pinning and bifurcations by GNPs (Li et al. 2008).

Sui et al. report a study that the concepts of "liquid nano-reinforcements" and hybrids of CNFs and CNTs are introduced to realize the improvement of reinforcing and toughening epoxy resins simultaneously. Results show that the flexural strength and breaking strain of the epoxy composites with 0.3 wt% reactive carbon nanofibers (CNFs) and 0.2 wt% carbon nanotubes (CNTs) increased by 45 and 64 %, respectively, exceeding that of the epoxy nanocomposites singly reinforced by the same (0.5 wt%) content of CNFs. Dynamic mechanical analysis showed there was an obvious increase in storage modulus and glass transition temperature for the hybrid nanocomposites compared to that of the neat epoxy resin (Sui et al. 2009).

Shokrieh et al. investigated the effects of adding a combination of synthesized grapheme nanosheets and carbon nanofibers (CNFs) on the flexural fatigue behavior of epoxy polymer (Shokrieh et al. 2014). In this study, The flexural bending fatigue life of 0.5 wt% of graphene/CNF/epoxy hybrid nanocomposites has been considered at room temperature. The samples were subjected to different displacement amplitudes fatigue loadings. Due to the addition of hybrid nanoparticles, a remarkable improvement in fatigue life of epoxy resin was observed in comparison with results obtained by adding 0.25 wt% graphene or 0.25 wt% CNF into the resin. Experimental observations show that at a strength ratio equal to 43 % by using 0.5 wt% of hybrid nanoparticles; 37.3-fold improvement in flexural bending fatigue life of the neat epoxy was observed. While, enhancement of adding only graphene or CNF was 27.4 and 24-fold, respectively. The flexural strength hand stiffness in this experimental research were presented in Table 4.8.

Results show that the GPL increased the stiffness of nanocomposites and the pull-out of the CNT increases the strength. Therefore, hybridization of these nanoparticles promotes the toughening and influences the crack propagation characteristics of the pure polymer which in turn causes a significant improvement in fatigue life of nanocomposites (Shokrieh et al. 2014).

Ayatollahi et al. studied the effect of simultaneous presence of nanoclay and MWCNT on the electrical and mechanical prosperities of the resulting nanocomposite (Ayatollahi et al. 2011e). The results of tensile tests in this research are depicted in Fig. 4.5. The addition of nanoparticles of any type generally increases the Young's modulus. According to this figure, the nanocomposite reinforced with 0.5 wt% MWCNT has higher modulus compared with that enhanced with 5 wt% nanoclay, which means that lower contents of MWCNT can lead to a better Young's modulus compared with nanoclay. Concurrent presence of 0.25 wt% MWCNT and 2.5 wt% nanoclay led to an increase in the Young's modulus similar to nanocomposites containing 5 wt% nanoclay. However, addition of 0.5 wt% MWCNT and 5 wt% nanoclay increased the Young's modulus dramatically compared with the pure epoxy. Indeed, the increase in the contents of nanofillers reduces the flexibility of the polymer chains which results in higher Young's moduli.

Table 4.8 Different properties obtained for: pure epoxy resin, 0.25 wt% GPL, 0.25 wt% CNF and 0.25 wt% of GPL plus 0.25 wt% of CNF hybrid nanoparticles epoxy nanocomposites, adopted from (Shokrieh et al. 2014)

Nanoparticles content	Flexural strength (MPa)	Flexural stiffness (GPa)
Pure epoxy resin	110	3.00
0.25 wt% CNF	121	3.18
0.25 wt% GPL	118	3.40
0.25 wt% of GPL + 0.25 wt% of CNF (hybrid)	123	3.43

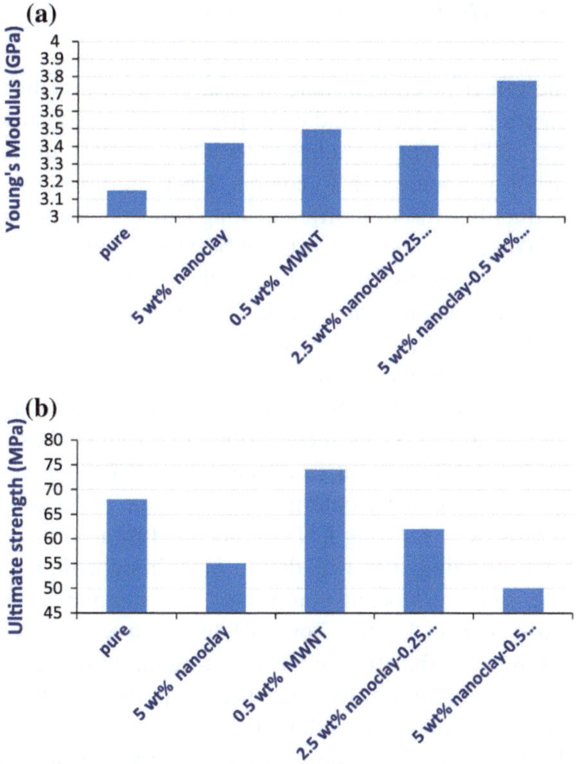

Fig. 4.5 The tensile properties of pure epoxy and hybrid nanocomposites **a** Young's modulus **b** tensile properties (Ayatollahi et al. 2011e)

However, the tensile strength of all types of nanocomposites decreased except for those reinforced with 0.5 wt% of MWCNT, which shows an increase around 7 %. Using nanoclay either as single filler or along with MWCNT generally leads to decrease in tensile strength. Since the increase in tensile strength has been reported before for the nanocomposite filled with small contents of nanoclay (Shokrieh et al. 2010), the reduction of tensile strength of the nanocomposite reinforced with high contents of nanoclay can be attributed to the agglomeration of nanofillers (Ayatollahi et al. 2011e).

4.4 Fracture Toughness

The addition of CNTs to the resin has two main effects: an increase of the neat resin elastic modulus and the introduction of toughening mechanisms such as CNT bridging. Depending on the embedded length of CNTs in the resin, CNT bridging can involve either CNT pull-out or CNT rupture. In both cases, the nanotubes bridge the crack surfaces shielding the crack front from carrying the entire tensile load. In this section, fracture toughness of reinforced nanocomposites by CNTs will be studied.

4.4.1 Mode I Fracture Toughness

The effect of carbon nanotubes as the reinforcement of the matrix in composite materials has been studied experimentally (Thostenson and Chou 2006; Seyhan et al. 2009). These studies showed an increase in fracture toughness even at low-carbon nanotube (CNT) content. CNT reinforced resins can increase the composite ductility through different toughening mechanisms such as CNT pull-out (i.e. CNT/matrix debonding), CNT bridging, and crack deviation (Gojny et al. 2005; Faber and Evans 1983). Among the observed mechanisms, CNTs bridging is the only mechanism that benefits from the high CNTs mechanical properties. It was shown that crack deviation is mostly controlled by the shape of the reinforcing phase (Faber and Evans 1983). More recently, the fracture toughness of multi-walled CNT (MWCNT) modified polymer was modelled (Blanco et al. 2009; Tong et al. 2008). In these works, two toughening mechanisms were considered: the MWCNT pull-out from the matrix and a sword-in-sheath mechanism caused by the failure of the outermost layer of the MWCNT.

However, for long CNTs embedded in a polymer, there is a critical length for CNT bridging that will define other toughening mechanisms. By analogy with long fibre reinforced composites, the nanotubes will pull-out if their length is below a critical value. For CNT having a length higher than a critical value, there will be a combination of CNT pull-out and rupture (Hull 1981).

Inconsistent results have been reported on how different methods of material processing and different types of CNTs affect mode I fracture toughness (K_{Ic}) of nanocomposites. Various methods of dispersing nanotubes in resin, such as stirring, calendaring and sonication, have also been reported in the literature. For instance, Thostenson and Chou used the calendaring technique to disperse MWCNTs in epoxy and used single-edge notch bend (SENB) specimen to measure K_{Ic}. According to this study, maximum K_{Ic} was obtained for 0.2 wt% of MWCNTs when the gap between the rolls of calendaring machine was 10 μm (Thostenson and Chou 2006). However, Zhou et al. reported 17, 24, 30 and 4 % increase in K_{Ic} by adding 0.1, 0.2, 0.3 and 0.4 wt% of MWCNTs to the epoxy using sonication (Zhou et al. 2008).

Moreover, the effects of functionalization of CNTs on mechanical properties including K_{Ic} were investigated in some studies. For example, Sun et al. added 1 wt% of pristine and functionalized single wall carbon nanotubes to epoxy resin using sonication technique. Fracture toughness increased 3 and 18 % by addition of pristine and functionalized single wall carbon nanotube, respectively (Sun et al. 2008). Ma et al. used 3-Glycidoxypropyltrimethoxysilane as the functionalization agent for MWCNTs. They used sonication to disperse MWCNTs and the CT specimen to measure K_{Ic}. They added 0.05, 0.1, 0.25 and 0.5 wt% of both pristine and functionalized MWCNTs to epoxy. Fracture toughness decreased 2.5, 15, 23 and 24 % by addition of untreated MWCNTs and increased 2.5, 6, 9 and 10 % by addition of functionalized MWCNTs, respectively (Ma et al. 2007).

Mirjalali et al. modelled the effect of aligned and randomly oriented carbon nanotube (CNT), with respect to the crack growth plane, on the mode I fracture toughness of polymers and finally, experimental results are compared with the model predictions. Both the experimental results and the model show that for randomly dispersed nanotubes with low volume fraction (lower than 2 %), no significant enhancement in toughness can be achieved. Therefore, toughness enhancement with CNTs requires increasing the volume fraction and length of CNTs, and aligning them normal to the crack growth plane. These modifications are bounded by the limitations and challenges in the processing of the CNT polymer mixture (Mirjalili and Hubert 2010).

Based on the developed theory, CNTs have the potential to improve the fracture toughness of brittle resins in mode-I fracture. The results also suggest that the CNT rupture is the main toughening mechanism in CNT-modified resins. Accordingly, to improve the fracture toughness of CNT modified polymers, long CNTs with high volume fraction should be incorporated into the resin system, aligned perpendicular to the fracture growth surface. Figure 4.6 shows the recommended steps to improve the fracture toughness of CNT-modified resins (Mirjalili and Hubert 2010).

Based on the CNT bridging model presented in this study, aligning relatively long carbon nanotubes perpendicular to the crack growth plane has great potential to enhance the toughness of brittle polymers. This improvement is due to the toughening mechanisms, mainly CNT bridging, that these nano particles introduce inside a polymer. The model also shows that in most cases SWCNTs were the best choice for toughening. However, according to the geometry of the CNTs, the critical length may vary, resulting to the fact that MWCNT becomes the optimum choice (Mirjalili and Hubert 2010).

Ayatollahi et al. studied the effect of adding nanoclay and MWCNT on mode I fracture prosperities of nanocomposites. In this research, In order to find out the proper content of filler for obtaining better fracture resistance, a number of nanocomposites based on different amounts of filler were prepared and tested. The results are presented in Table 4.9. For epoxy/MWCNT nanocomposite the maximum fracture toughness was obtained at 0.5 wt% of filler with 26.3 % increase

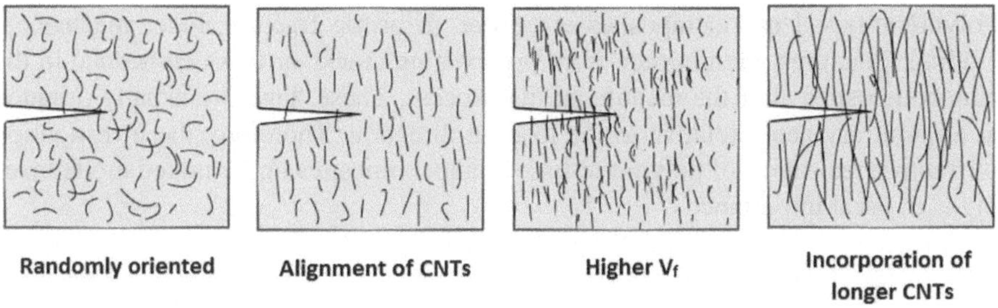

Fig. 4.6 Different recommendations for improving the toughness of brittle polymers by incorporating CNTs (Mirjalili and Hubert 2010)

Table 4.9 Fracture toughness of hybrid nanocomposites reinforced with MWCNTs and nanoclay at different contents (Ayatollahi et al. 2011e)

Sample	MWCNT (wt%)	Nanoclay (wt%)	Fracture toughness (MPa $m^{0.5}$)	Increase in K_{Ic} (*compared with pure epoxy*)
Epoxy	–	–	1.62	–
Epoxy/MWCNT	0.1	–	1.87	15.2 %
Epoxy/MWCNT	0.5	–	2.04	26.3 %
Epoxy/MWCNT	1	–	1.93	19.2 %
Epoxy/nanoclay	–	1	1.81	11.7 %
Epoxy/nanoclay	–	2	1.87	15.5 %
Epoxy/nanoclay	–	3	1.90	17.2 %
Epoxy/nanoclay	–	4	1.98	22.3 %
Epoxy/nanoclay	–	5	2.03	25.5 %

compared with pure epoxy. For epoxy/nanoclay nanocomposites the maximum fracture toughness was obtained at 5 wt% amount of filler with 25.5 % increase (Ayatollahi et al. 2011e).

Taking the experimental results into account, two different combinations of MWCNT and nanoclay namely 0.5 wt% MWCNT-5 wt% nanoclay and 0.25 wt% MWCNT-2.5 wt% nanoclay were chosen to study the effect of simultaneous presence of both fillers on the fracture behavior of the resulted nanocomposites. The results are presented in Fig. 4.7. As can be seen, the fracture toughness of nanocomposite enhanced with 0.25 wt% MWCNT-2.5 wt% nanoclay is close to the maximum fracture toughness obtained for the nanocomposite reinforced with a single filler. In addition, the nanocomposite containing 0.5 wt% MWCNT-5 wt% nanoclay shows the highest amount of K_{Ic}. The results show that the concurrent presence of MWCNTs and nanoclay has a positive effect on increasing the toughness of nanocomposites.

4.4.2 Mixed Mode Fracture Toughness

Polymeric-based nanocomposites are prone to brittle fracture under mechanical loading. Fractures often initiate from pre-existing cracks or flaws embedded in the nanocomposites after the manufacturing process. These flaws are usually oriented at an arbitrary angle relative to the loading direction. Therefore, the investigation of fracture resistance of polymeric-based nanocomposites under mixed mode loading is of great importance.

A review of the papers published on fracture toughness of polymer/CNT nanocomposites shows that mostly those studies are limited only to mode I fracture. Indeed, mixed mode fracture in polymers has been studied but only for pure materials and not nanocomposites (Ayatollahi and Aliha 2009; Aliha and Ayatollahi 2010). Some researchers have used different fracture criteria to estimate mixed

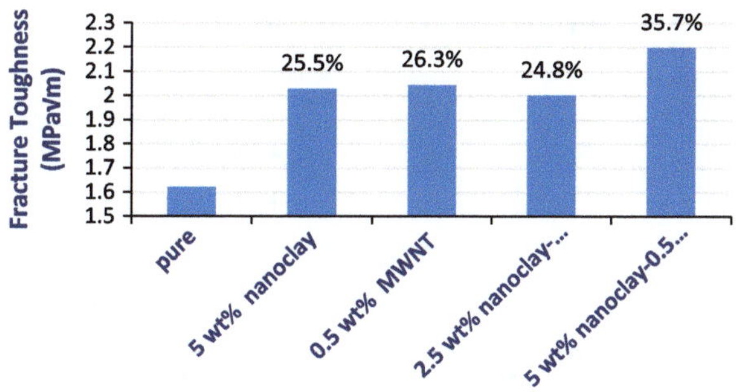

Fig. 4.7 Fracture toughness of hybrid nanocomposites reinforced with different types of fillers (Ayatollahi et al. 2011e)

mode fracture toughness of polymers. For instance, the brittle fracture behavior of PMMA under mixed mode I/II loading was investigated by Maccagno et al. where the experimental results were used to evaluate the maximum tangential stress (MTS), minimum strain energy density (SED) and maximum energy release rate (G) criteria,with the first criterion showing the best agreement (Maccagno and Knott 1989). Moreover, in studies worked by Ayatollahi, a generalized maximum tangential stress criterion was employed for analyzing mixed mode fracture in PMMA and dental resin materials (Ayatollahi et al. 2006; Aliha and Ayatollahi 2008). Araki et al. also investigated the fracture initiation and fracture angle for epoxy resin at room temperature under mixed mode loading and examined four fracture criteria namely, MTS, SED, G and an empirical criterion to estimate the experimental results. They concluded that the common mixed mode fracture criteria could not be applied to their investigated epoxy resin (Araki et al. 2005).

Ayatollahi et al. studied the effects of MWCNTs on fracture toughness of epoxy/MWCNTs nanocomposites under combined shear and normal loading (Ayatollahi et al. 2011f). In this research The FE simulation is used to determine the appropriate boundary conditions for mode II and mixed mode loading. Using FE results, two equations were derived for calculating fracture toughness from the fracture load under mixed mode loading conditions. Afterwards, epoxy/MWCNT nanocomposites containing 0.1, 0.5 and 1 wt% of MWCNTs were prepared using the sonication technique. Then, pure mode I, pure mode II and mixed mode fracture toughness of nanocomposite were experimentally measured and the effects of mode mixity on its effective fracture toughness $K_{eff} = (K_I^2 + K_{II}^2)^{0.5}$ were studied (Ayatollahi et al. 2011f).

The notched three-point bend specimen (Fig. 4.8a) was employed for mode I fracture experiments and the single-edge notch bend specimen under four-point bend (FPB) loading (see Fig. 4.8b) was used for determining mixed mode fracture resistance of the epoxy/MWCNT nanocomposite. When the FPB specimen is subjected to anti-symmetric loading (i.e. $L_1 = L_4$ and $L_2 = L_3$), bending moment vanishes along the crack plane and there is only a shear force at the crack position. On the other hand, different combinations of L_1/L_4 and L_2/L_3 leads to different amounts of shear force and bending moment along the crack plane covering a wide range of mode mixity. Consequently, the four-point bend (FPB) specimen

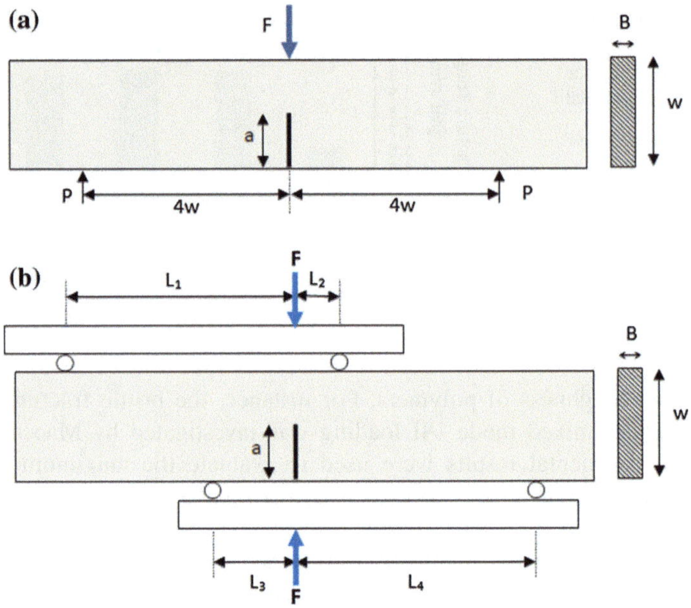

Fig. 4.8 The experimental set up **a** three-point bend specimen (pure mode I) **b** four-point bend specimen (mixed mode and mode II) (Ayatollahi et al. 2011f)

is often considered as a suitable specimen for mixed mode fracture tests provided appropriate values are used for the dimensions L_1, L_2, L_3 and L_4 shown in Fig. 4.8b.

Mode I fracture toughness (K_{Ic}) was determined using the three-point bend specimens according to ASTM-D5045–99. Mode II and mixed mode fracture toughness of nanocomposite was measured using a $10 \times 20 \times 110$ mm notched FPB specimen. To prepare the specimens for fracture tests, first, a primary crack was created in the specimens using a very thin saw blade. Then, the crack tip was sharpened utilizing a razor blade.

The force-displacement diagrams for all the fractured specimens including pure mode I, pure mode II and mixed mode specimens were linear up to the fracture load. Typical force-displacement diagrams are shown in Fig. 4.9 for mode I fracture tests conducted on pure epoxy and nanocomposites containing different amounts of filler.

While, the mode I specimens fractured all along the direction of initial crack, the crack extension in mixed mode and mode II specimens took place at angles between 0° and 70° relative to the initial crack. Figure 4.10 shows samples of nanocomposite specimens broken under mode I and mode II loading conditions.

The mixed mode fracture toughness results are shown in Fig. 4.11. According to this figure, the fracture resistance under all mode mixities increases with increasing MWCNT contents up to 0.5 wt%. However, the nanocomposite containing 1 wt% of MWCNT shows a different behavior compared with the one

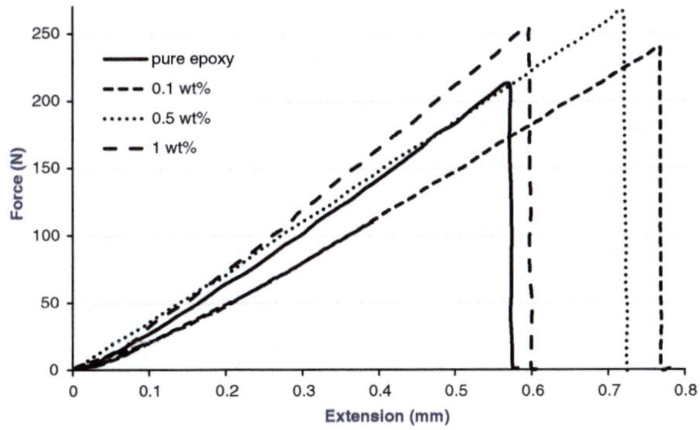

Fig. 4.9 Typical force-displacement curves obtained for pure epoxy and for nanocomposites under mode I loading conditions (Ayatollahi et al. 2011f)

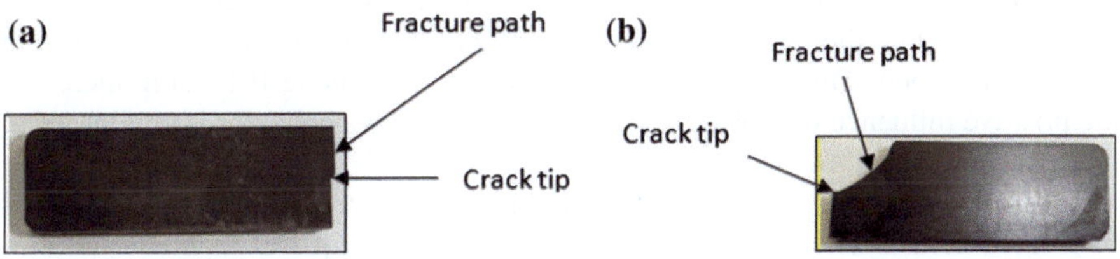

Fig. 4.10 Samples of epoxy/0.1 wt% nanocomposite broken under **a** mode I **b** mode II loading conditions (Ayatollahi et al. 2011f)

Fig. 4.11 Mixed mode I/II fracture toughness data obtained for pure epoxy and for nanocomposites (Ayatollahi et al. 2011f)

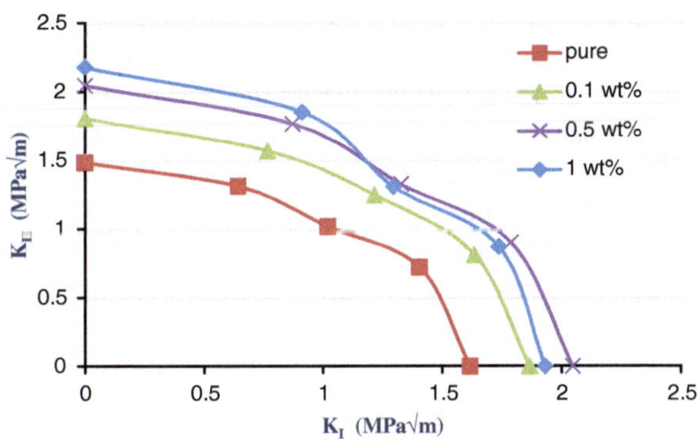

containing 0.5 wt%. Indeed, for mode I dominant loading the nanocomposite containing 0.5 wt% of filler exhibits a higher fracture toughness compared to the one with 1 wt%. By moving towards mode II, the difference between the curves of 0.5 and 1 wt% MWCNT reduces first, and eventually for mode II dominant loading conditions, the fracture resistance of the nanocomposite containing 1 wt% of MWCNT become higher than that containing 0.5 wt% of MWCNT.

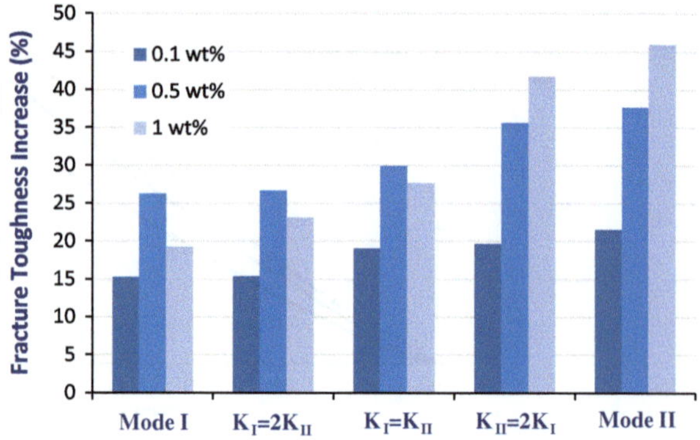

Fig. 4.12 Increase in effective fracture toughness of nanocomposites containing 0.1, 0.5 and 1 wt% MWCNTs, relative to pure epoxy for different combinations of mode I and mode II (Ayatollahi et al. 2011f)

Figure 4.12 shows the increase in the effective fracture toughness ($K_{eff} = (K_I^2 + K_{II}^2)^{0.5}$) of nanocomposite compared to pure epoxy for different modes mixities. It is seen that the increase in fracture toughness of nanocomposite depends on the loading mode. That is to say, as the dominancy of mode II loading increases, the positive influence of MWCNT on K_{eff} enhances. This behavior is similar for all three contents of MWCNT mixtures, 0.1, 0.5 and 1 wt%.

It is often suggested that mixed mode brittle fracture is due to the attainment of a fixed critical energy/tensile stress near the crack tip for all mode mixities. Therefore, the mechanisms such as crack deviation and crack bridging, which contribute in increasing fracture toughness under mode I loading conditions, are expected to participate in mixed mode and mode II loading conditions as well. Therefore, there must be some extra mechanisms which are responsible for different effects of MWCNTs on fracture toughness under different mode mixities. To identify these extra mechanisms, the micro-mechanisms involved in the process of fracture under mixed mode and mode II loading should be studied. Some of these mechanisms were investigated by Bhattacharjee and Knott (1995). For instance, the presence of higher stresses ahead of the crack tip in shear loading can be one reason for the misty areas which appear in the fracture surface under mixed mode and mode II fracture surface. In other words at some stages of crack growth, the stress becomes high enough to initiate secondary local cracks (Bhattacharjee and Knott 1995), which give rise to the rougher area next to the crack front. The creation of secondary cracks due to higher stresses along the crack front under shear loading can provide an extra opportunity for MWCNTs to dissipate more energy because more local crack nucleation requires higher fracture energy particularly in the presence of MWCNTs. Another reason for higher roughness of surfaces under shear loading is the fact that in shear loading unlike normal loading, fracture does not take place along the initial pre-crack. In fact, the fracture initiate at angle θ_0 which increases with increasing the mode II component (Ayatollahi et al. 2011f).

One can suggest that the third toughening mechanism occurs because once fracture initiates at an angle θ_0, it does not grow in a straight line and the crack path profile is curved (as shown in Fig. 4.10b). The curved path of crack growth can be due to the fact that the maximum tangential stress direction changes as the crack advances. Thus, at each step of crack propagation, the crack plane rotates and grows in a new direction which has the maximum tangential stress. The presence of stiff MWCNTs in each step of crack plane rotation can provide a high resistance against crack propagation, which results in a higher fracture energy. As declared earlier, both deep cleavages and shallow and dendriform cracks were observed as the shear component of loading increased. The deep cleavages observed in both pure and enhanced epoxy can be due to the crack plane rotation, and the shallow cleavages might be because of both the creation of secondary cracks and the crack deviation mechanisms discussed for mode I fracture (Ayatollahi et al. 2011f).

All the mechanisms suggested above have greater effects as the shear component of loading increases and they are in addition to those mechanisms involved in mode I fracture. Thus, they can be used to justify the increase in fracture toughness when the mode II component of load increases, as shown in Fig. 4.11. It was stated earlier that under mode I loading, fracture toughness of the nanocomposites containing 1 wt% MWCNT was lower than those containing 0.5 wt% MWCNT, which was attributed to the aggregates observed in the former. As the shear component of loading increased, the fracture toughness of nanocomposites containing 1 wt% MWCNT enhanced such that near mode II loading it was higher than those containing 0.5 wt% MWCNT. One can suggest two reasons for this trend. First, this opposing behavior might be due to the additional energy dissipating mechanisms that are presented under shear loading compared to mode I loading, which outweigh the negative effects of agglomeration. The other explanation can be the fact that the relatively larger agglomerates are able to interact with the crack front more effectively under shear loading. It implies that the presence of aggregates may be useful to some extent when the shear component of loading is dominant (Ayatollahi et al. 2011f).

Ayatollahi et al. studied the effects of multi-walled carbon nanotubes (MWCNTs) on the mechanical properties of epoxy/MWCNT nano-composites with emphasis on fracture toughness under bending and shear loading conditions. It is seen that the increase in fracture toughness of nano-composite depends on the type of loading. That is to say, the presence of MWCNTs had a greater effect on fracture toughness of nano-composites under shear loading compared with normal loading (Ayatollahi et al. 2011c). In this study, K_{Ic} and K_{IIc} were determined and the obtained results are presented in Table 4.10. According to Table 4.10, with addition of MWCNT to epoxy, both K_{Ic} and K_{IIc} increase.

Table 4.10 Mode I and mode II fracture toughness obtained for pure epoxy and for epoxy/MWCNT nano-composites (Ayatollahi et al. 2011c)

Mode of fracture	Pure epoxy	0.1 wt% MWCNT	0.5 wt% MWCNT	1 wt% MWCNT
Mode I (MPa $m^{0.5}$)	1.619 ± 0.14	1.865 ± 0.07	2.045 ± 0.1	1.93 ± 0.19
Mode II (MPa $m^{0.5}$)	1.49 ± 0.15	1.815 ± 0.18	2.055 ± 0.2	2.18 ± 0.25

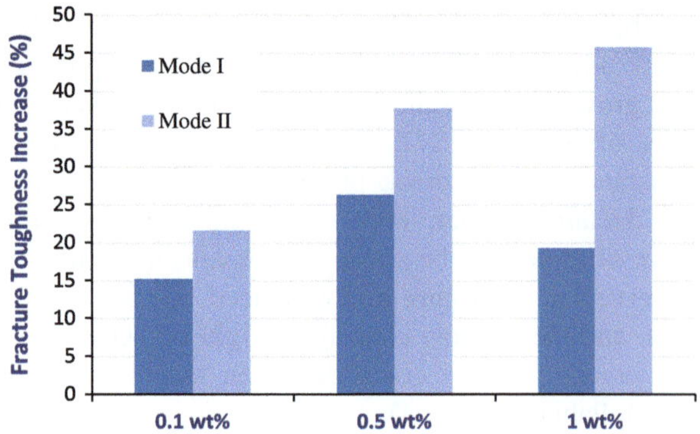

Fig. 4.13 The increase in mode I and mode II fracture toughness in different percent contents of MWCNT (Ayatollahi et al. 2011c)

Figure 4.13 shows the values of increases in K_{Ic} and K_{IIc} of the nano-composite compared to the pure epoxy for different amounts of filler. Two important points can be inferred from Fig. 4.13. First, regardless of the amount of filler, the presence of MWCNT causes more increase in fracture toughness of epoxy under shear loading compared with mode I loading. The second point is that in mode I loading, fracture toughness of the nano-composite containing 1 wt% MWCNT is lower than that containing 0.5 wt% MWCNT while in shear loading the contrary occurs.

The results of this study showed that the mode I fracture toughness of nanocomposites containing 1 wt% MWCNT was lower than those containing 0.5 wt% MWCNT which was attributed to aggregates observed in the fracture surfaces of all specimens using SEM and optical microscopy. However, the opposite happened in shear loading conditions. This opposing behavior might be due to the extra energy dissipating mechanisms under shear loading compared to fracture under mode I loading, which outweigh the negative effects of agglomeration. Moreover, a possible reason for the mentioned trend might be the fact that larger agglomerates are able to interact with crack front more effectively under shear loading (Ayatollahi et al. 2011c).

Ayatollahi et al. studied mixed mode fracture resistance of epoxy-based nanocomposites reinforced with carbon nanoparticles of three different shapes. The nanodiamond (ND) of spherical shape, the carbon nanofiber (CNF) of cylindrical shape and the graphene oxide (GO) nanoplatelets are used to prepare the nanocomposites. The semi-circular bending specimen (SCB) is employed to perform a total number of 36 mixed mode fracture tests on the prepared nanocomposites. Fracture toughness of nanocomposites having high aspect ratio fillers (i.e. CNF and GO) is found to be higher than that of ND nanocomposites (Shadlou et al. 2013).

In this study, according to the experimental results depicted in Fig. 4.14, the nanoreinforcements lead to considerable enhancement in mixed mode fracture resistance of epoxy. One of the most important fracture mechanisms in nanocomposites is crack deflection (Wetzel et al. 2006). Based on this mechanism, when the crack growth path reaches a nanoparticle deviates from its initial path and propagates along the interface area between the particle and the matrix. Hence, the

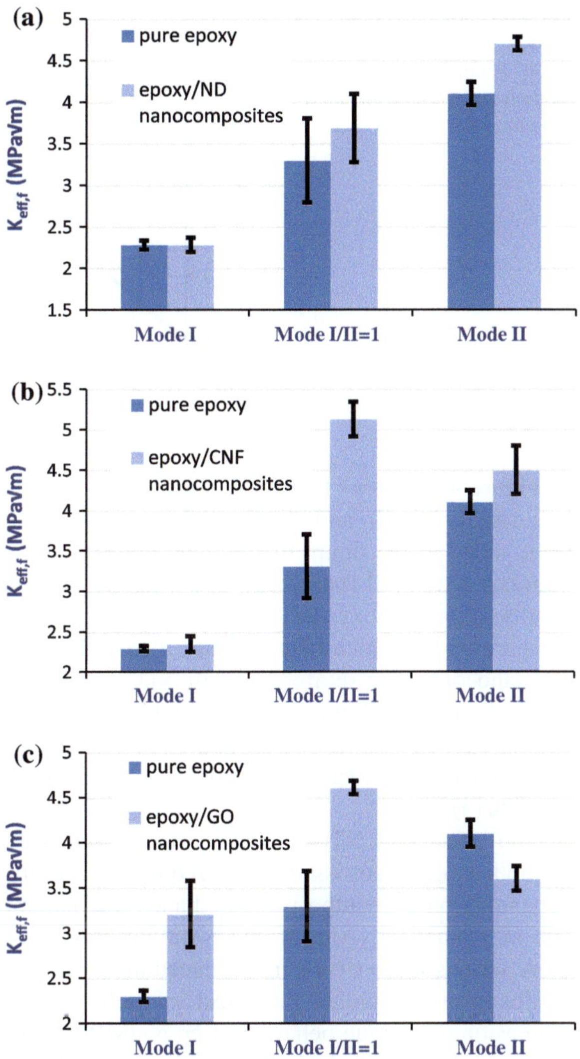

Fig. 4.14 Effective fracture resistance $K_{eff\,f}$ for pure epoxy and for nanocomposites reinforced with different types of nanoparticles **a** ND, **b** CNF and **c** GO (Shadlou et al. 2013)

crack propagates in a longer path and its extension requires higher fracture energy. Meanwhile, the path of crack growth in mixed mode brittle fracture is often curved. A curved fracture trajectory comes across more nanoparticles during its extension resulting in an increased reinforcement efficiency of these particles. The size of nanoreinforcements can also affect the micromechanism of crack growth. Larger particles create longer crack growth path and dissipate more energy. In conclusion, the cylindrical-shape CNF and the GO platelets that are larger than the spherical-shape ND particles, have more tendencies to improve the fracture resistance of nanocomposites (Shadlou et al. 2013).

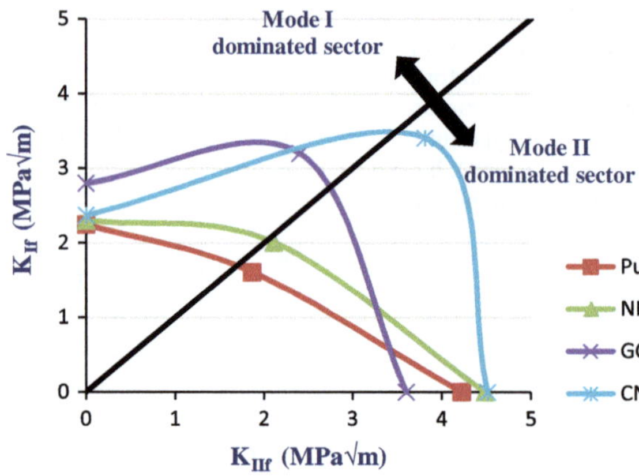

Fig. 4.15 The curves of K_{If}-K_{IIf} plotted based on the results obtained from mixed mode fracture experiments (Shadlou et al. 2013)

Figure 4.15 shows the relation between the critical mode I and mode II stress intensity factors K_{If} and K_{IIf} for different mode mixities. The bisector line has also been added to the figure to show the dominated mode I (above the line) and the dominated mode II (below the line) sectors. According to Fig. 4.15, the trends of results for pure epoxy and for nanocomposites are in some extent different. Unlike the pure epoxy which demonstrates a smooth decrease in K_{If} by increasing K_{IIf}, the nanocomposite reinforced with ND particles has almost a constant value of K_{If} in the mode I dominated sector while the CNF and GO nanocomposites have increasing trends in this sector. The reason for this difference may be related to local constrains which generated due to the presence of nanoparticles (Shadlou et al. 2013).

The effect of CNT dimensions on the fracture properties of nanocomposites is another important issue which requires a careful investigation. This is because depending on the synthesis method and other parameters involved, the CNT diameter may practically vary from several nm to several hundred nm. This problem is more serious when CNTs are produced industrially in large quantities.

Zhang et al. studied the effect of multi-walled carbon nanotube (MWCNT) dimension on fatigue behavior of epoxy/MWCNTs nanocomposites. Their results indicated that longer MWCNTs with smaller diameters (i.e. higher aspect ratios) were well suited for the reduction of fatigue crack growth rate (Zhang et al. 2008).

Ayatollahi et al. used four types of MWCNTs with different diameters for investigating the effects of multi-walled carbon nanotube (MWCNT) diameter on fracture behavior of epoxy based nanocomposites under mixed mode I/II loading. They found that the aspect ratio of filler has huge impact on the fracture behavior of nanocomposite (Ayatollahi et al. 2011g).

In this study, different diameters of MWCNTs namely $20 < d < 30$, $30 < d < 40$, $40 < d < 50$ and $50 < d < 60$ nm and the intrinsic length of 10–30 μm as fillers were used. These four types of nanocomposites will be referred to hereafter as type A, type B, type C and type D, respectively. Figure 4.16 shows the results obtained from the fracture tests on a K_{II} versus K_I diagram. As can be seen, the

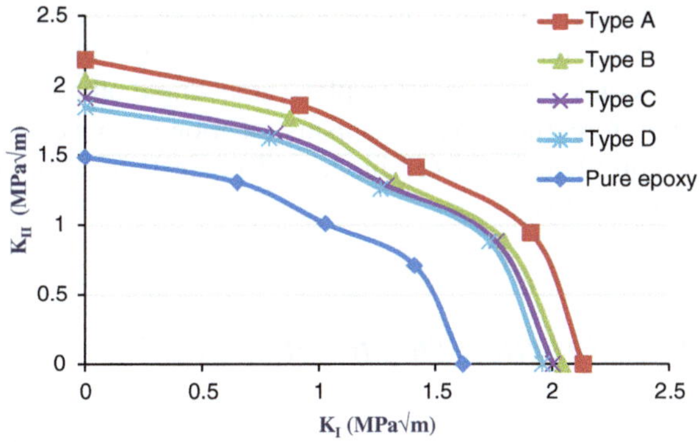

Fig. 4.16 Effects of MWCNT aspect ratio on mixed mode I/II fracture toughness data obtained for pure epoxy and for nanocomposites (Ayatollahi et al. 2011g)

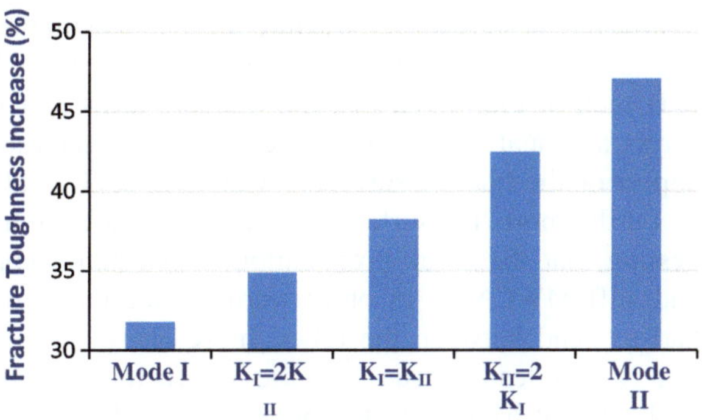

Fig. 4.17 Percent increase in effective fracture toughness of nanocomposite containing type A for different mode mixities (Ayatollahi et al. 2011g)

fracture toughness of nanocomposite is generally higher than that of pure epoxy. In addition, for all mode mixities, the fracture toughness of nanocomposite increases as the aspect ratio of the fillers grows. Figure 4.17 shows the increase in the effective fracture toughness ($K_{eff} = (K_I^2 + K_{II}^2)^{0.5}$) of epoxy/type A MWCNT nanocomposite compared with that of pure epoxy for different mode mixities. It is seen that the increase in fracture toughness of nanocomposite depends on the relative combination of mode I and mode II. In other words, as the dominancy of mode II loading increases, the positive influence of MWCNT on K_{eff} enhances.

In order to explain the reason for the effect of MWCNT diameter on fracture toughness of nanocomposites, the parameters which could be affected by filler diameter should be studied. Since the load transfer from matrix to MWCNT ensures the high efficiency of composite system, the interfacial strength between them is of crucial importance. Several researchers have shown that as the diameter of cylindrical particles reduces the interface strength increases. Stronger interface between the epoxy and the stiff MWCNT leads to higher mechanical properties (Barber et al. 2004).

Two well-known mechanisms which can play important roles in mode I fracture toughness are crack deviation and crack bridging. However, mixed mode brittle fracture is due to the attainment of a fixed critical energy/tensile stress near the crack tip for all mode mixities. The major distinction in mixed mode fracture is that crack extension takes place at an angle other that the direction of initial crack. Therefore, the two mechanisms mentioned above for mode I fracture are expected to contribute in mixed mode and mode II loading conditions as well (Ayatollahi et al. 2011g).

4.4.3 Fracture Surface Studies

The fracture surfaces of specimens could be studied using SEM and optical microscopy to recognize the reasons for the fracture mechanisms in mode I and mode II loading. Ayatollahi et al. explore the fracture surfaces using SEM to recognize the reasons for the fracture mechanisms in the broken samples. Figure 4.18 shows typical pictures taken from specimen surfaces broken under mode I loading. It is seen in Fig. 4.18a that the fracture surface of neat epoxy is smooth, which shows a typical brittle fracture behavior. With addition of filler, regular shear steps appear on the fracture surfaces (Ayatollahi et al. 2011f).

Furthermore, the surface roughness increases in higher MWCNT contents suggesting that the crack propagation in the nanocomposite was opposed by rigid and stiff MWCNTs. In other words, since the segments of primary crack front had to bend between the tubes, more energy was needed for such local deviation which leads to an increase in fracture toughness. However, the addition of 1 wt% of MWCNTs to epoxy decreased K_{Ic} compared to the nanocomposite containing 0.5 wt% of MWCNTs. Besides, in the 1 wt%-MWCNT nanocomposite the regularity of shear steps is lost (see Fig. 4.18d) which can be due to MWCNT agglomerates. For a more detailed inspection, high magnification pictures were also taken from the fracture surfaces (Fig. 4.19) which revealed that in lower contents of MWCNTs (i.e. 0.1 and 0.5 wt%), fillers were dispersed very well (Fig. 4.19a, b, d). Meanwhile, as shown in Fig. 4.19c, in specimens containing 1.0 wt% of enhancers, MWCNTs tended to be agglomerated. However, a more detailed study of the observed agglomerations revealed that the resin had penetrated into and wet the MWCNTs accumulation. Figure 4.18d also shows that in the areas where MWCNTs are aggregated the regular pattern of steps are affected and the tail-like structures are formed. These agglomerations can be responsible for the reductions in the tensile strength and K_{Ic} of the nanocomposite with 1 wt% of MWCNTs which was mentioned earlier. In all the specimens, some MWCNTs were seen pulled out of the fracture surface (shown for example in Fig. 4.19d by white arrows) indicating a bridging mechanism. Generally, the number of pulled out MWCNTs in the accumulation parts is more than those in the well dispersed parts. This behavior can be due to a weaker interface between MWCNTs and polymer in these areas, which makes it easier for MWCNTs to come out of the polymer matrix. The pulled out MWCNTs in aggregated parts show that these

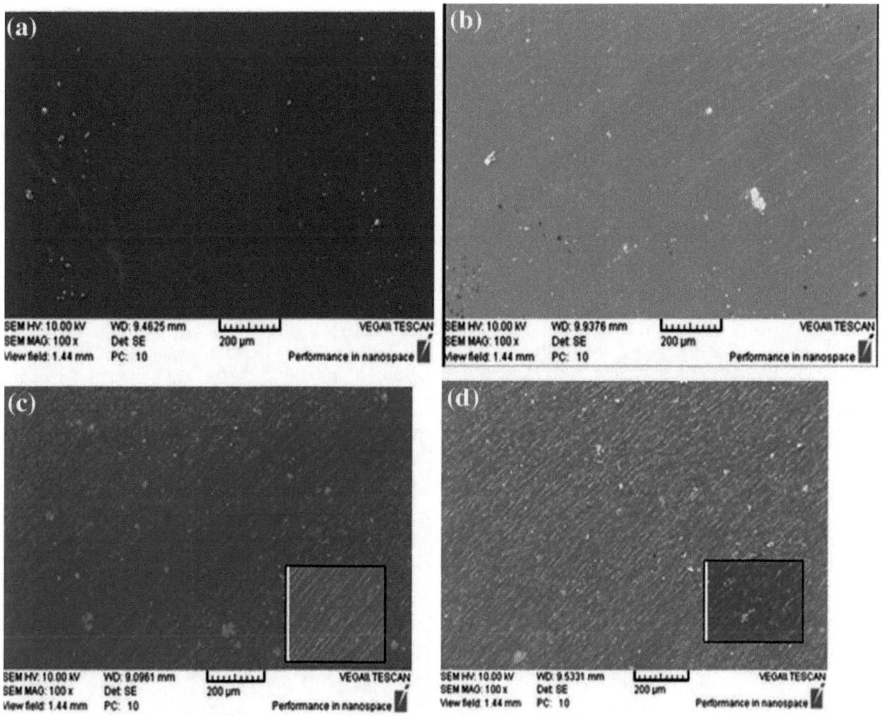

Fig. 4.18 Fracture surfaces for mode I fracture tests **a** neat epoxy **b** 0.1 wt% MWCNT/epoxy **c** 0.5 wt% MWCNT/epoxy **d** 1 wt% MWCNT/epoxy (in (**c**) and (**d**) higher magnifications are shown). All the pictures are taken from a region in the vicinity of the initial crack front. The direction of crack propagation: *top right* towards *bottom left* (Ayatollahi et al. 2011f)

agglomerations participate in toughening mechanisms and that they can be of different effects on the mechanical properties of nanocomposites.

In this research, the specimen broken under mixed mode and mode II loading conditions were also studied. Figure 4.20 shows the SEM micrographs of fracture surface for the pure epoxy and the nanocomposite containing 0.5 wt% of filler broken under mode I, mode II and mixed mode loading conditions. Two important points can be inferred from the taken pictures. First, in different modes of fracture the roughness of reinforced epoxy is generally higher than that of pure epoxy. The second point is that some deep cleavages are seen in both pure and enhanced epoxy surfaces. Besides, in nanocomposites, in addition to deep cleavages there are some shallow cleavages and dendriform cracks. With increasing the mode II contribution the density of these cleavages become higher. Thus, they are expected to be responsible for more increase in the toughness near mode II loading conditions, since higher cleavages can be considered to be associated with higher energy absorption. In addition, some changes in the fracture mechanisms were observed in mode II loading (Fig. 4.20c, f). For instance, some parabolic features

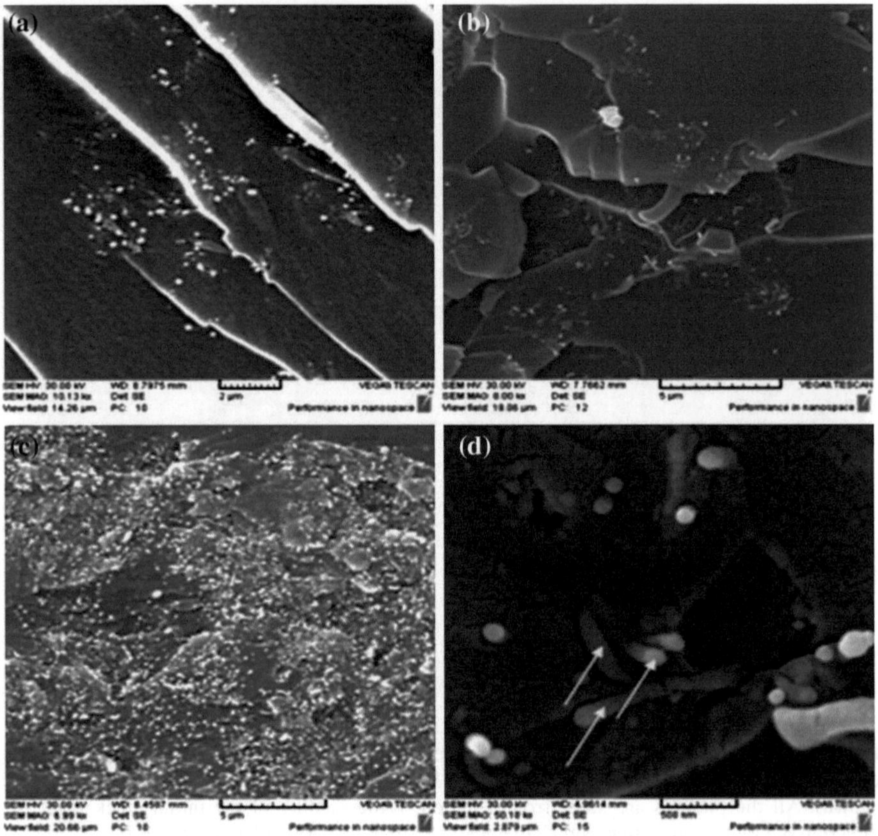

Fig. 4.19 Typical SEM micrographs fracture surfaces under mode I loading for epoxy/MWCNT nanocomposites **a** 0.1 wt%-well dispersed **b** 0.5 wt% well dispersed **c** an agglomerate in 1 wt% **d** pulled out MWCNTs in 0.5 wt% nanocomposite. All the pictures are taken from a region close to the initial crack front. The direction of crack propagation: *top right* towards *bottom left* (Ayatollahi et al. 2011f)

are present in pure epoxy specimens broken under mode II loading while for nano-composites there are no parabolic patterns. Since, this happened mainly in mode II loading conditions, the change in the cleavage patterns could be attributed to secondary cracks.

Ayatollahi et al. examined the fracture surfaces of the neat epoxy and the nano-composites using scanning electron microscopy (SEM) (Tescan VEGA-II SBU) and extracted some information related to the MWCNTs dispersion status and the fracture mechanisms. Figure 4.21 revealed that in lower contents of MWCNTs (i.e. 0.1 and 0.5 wt%) there is a well dispersion of fillers (Fig. 4.21a, b). Meanwhile, as shown in Fig. 4.21c, in specimens containing 1.0 wt% of enhancers, MWCNTs tended to be agglomerated (Ayatollahi et al. 2011c).

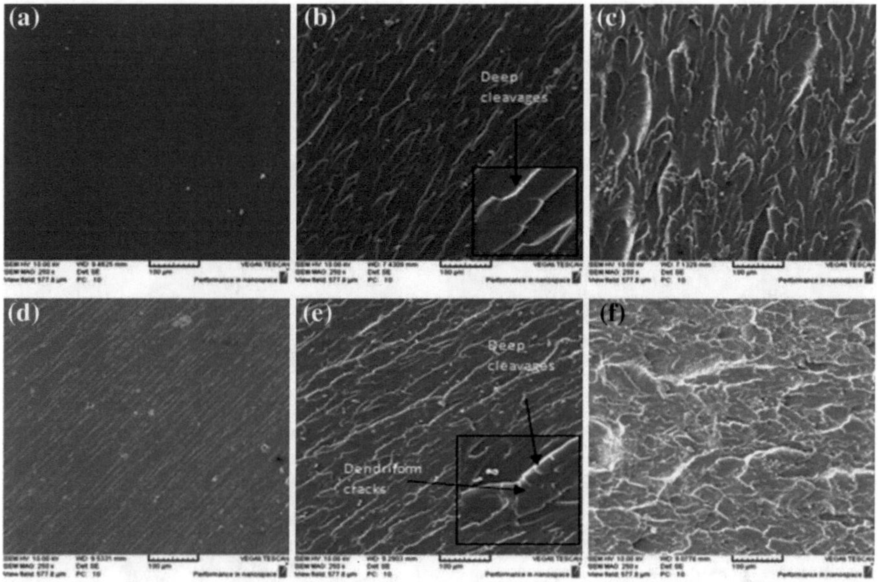

Fig. 4.20 Typical SEM micrographs of pure epoxy under **a** mode I **b** mixed mode ($K_I = K_{II}$) **c** mode II, and nanocomposites containing 0.5 wt% under **d** mode I **e** mixed mode ($K_I = K_{II}$) **f** mode II (in (**b**) and (**e**) higher magnifications are also shown). The pictures are taken from a region in the vicinity of the initial crack front. The direction of crack propagation: *top right* towards *bottom left* (Ayatollahi et al. 2011f)

Figure 4.22 shows the mode I and mode II fracture surfaces with higher magnification. It is seen that the mode I fracture surfaces (Fig. 4.22a, b) of both pure and enhanced epoxy are generally much smoother than the mode II fracture surfaces (Fig. 4.22c, d). While just some shallow scratches are seen in mode I fracture surfaces of epoxy/0.5 wt% MWCNT nano-composite, there are deep cleavages in both pure and enhanced epoxy surfaces related to mode II fracture. In nano-composites broken under mode II loading, there are also some shallow cleavages and dendriform cracks (Fig. 4.22d). The deep cleavages observed for both pure and enhanced epoxy in mode II specimens can be due to changing of crack plane.

4.5 Tribological Properties

The use of nano-particles in polymers for tribology performance enhancement started around mid-1990s and this area has become quite promising for the future as newer nanomaterials are being economically and routinely fabricated. It has also been reported that the gain in wear resistance for nanocomposite could also be attributed to an increase in the thermal conductivity of the composite compared to the pristine polymer. High thermal conductivity of the material facilitates a

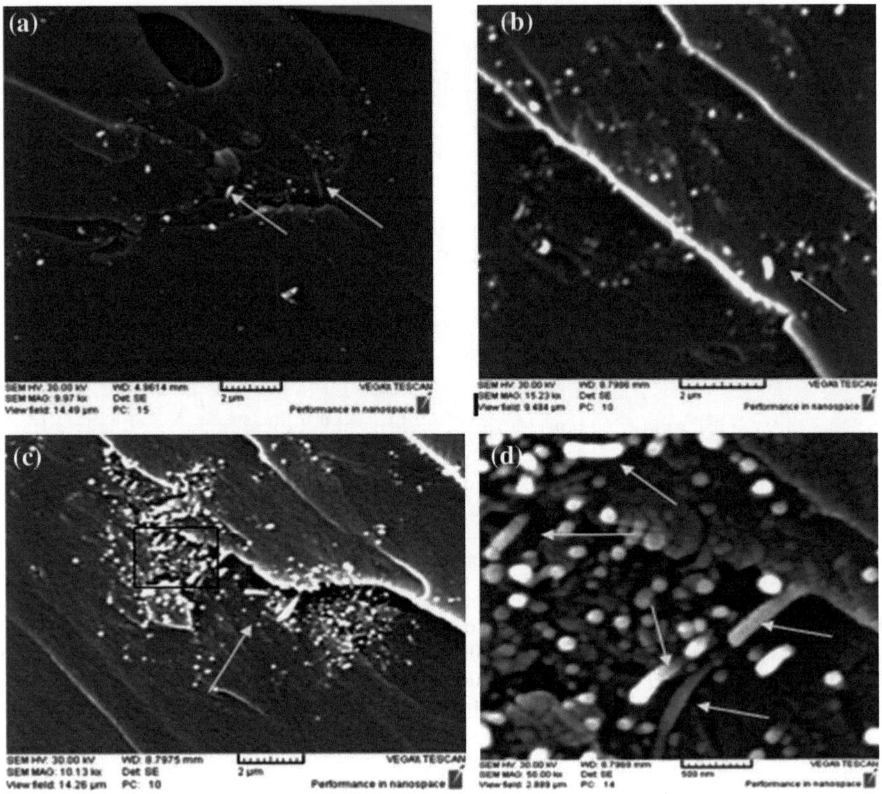

Fig. 4.21 Typical SEM micrographs of fracture surfaces for epoxy/MWCNT nano-composites: **a** 0.1 wt% **b** 0.5 wt% **c** 1 wt% **d** magnification of a selected part in (**c**) near the region of crack propagation. The direction of crack propagation: *top right* towards *bottom left* (Ayatollahi et al. 2011c)

lower mean interfacial temperature as the frictional heat generated at the interface is more readily transferred (conducted) to other parts of the machine (Brian and Sinha 2008).

CNT's role in a low-friction polymer, such as PTFE, is mainly of reinforcement type. Thus, the wear resistance is increased by at least three orders of magnitude with slight decrease in the coefficient of friction. Application of CNT as filler for UHMWPE and epoxy has similar effects on the wear resistance, however, in quantitative terms the improvement is only in the range of a few times to about an order of magnitude. For coefficient of friction, either there is marginal decrease in the value due to low real contact area owing to high load-bearing capacity of the nanocomposite, or there is slight increase in the coefficient of friction. The increase in the coefficient of friction is often believed to be because of the changes in the topography and due to the increased resistance of the CNTs to shearing actions at the interface (Brian and Sinha 2008).

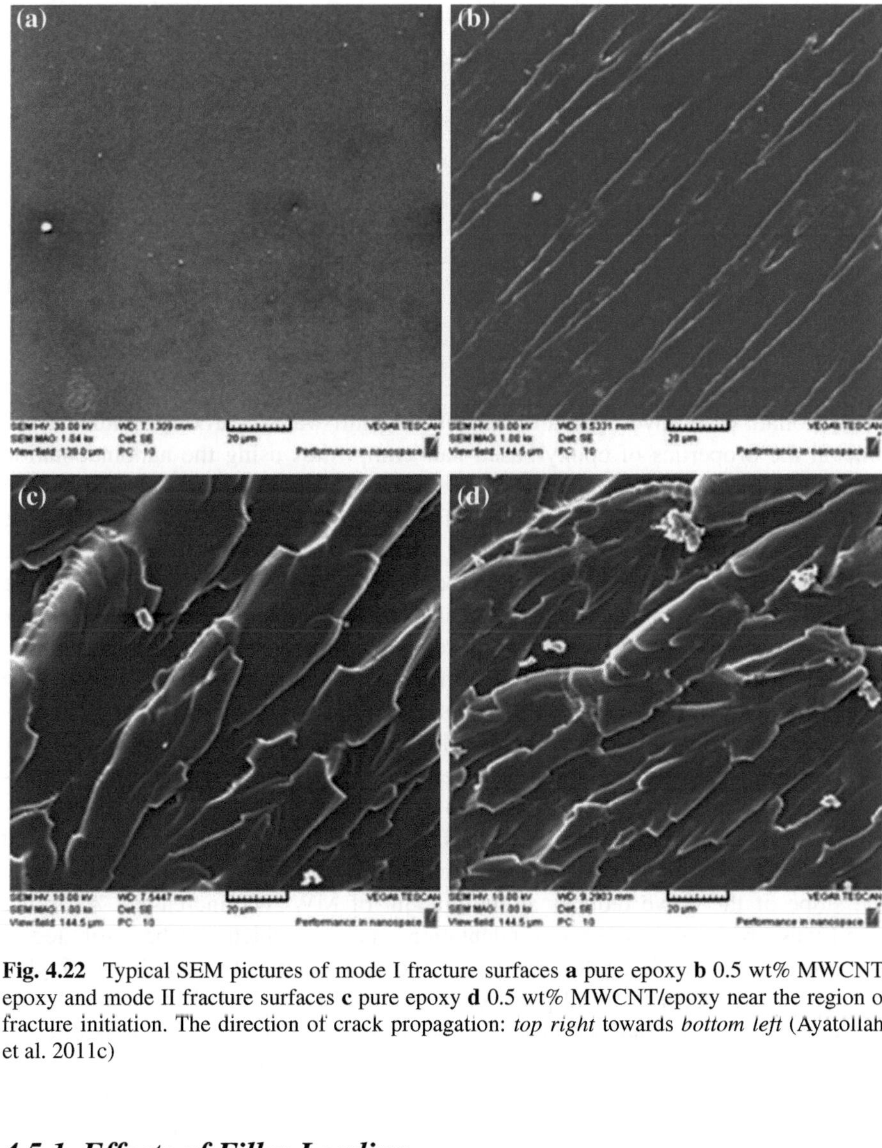

Fig. 4.22 Typical SEM pictures of mode I fracture surfaces **a** pure epoxy **b** 0.5 wt% MWCNT/epoxy and mode II fracture surfaces **c** pure epoxy **d** 0.5 wt% MWCNT/epoxy near the region of fracture initiation. The direction of crack propagation: *top right* towards *bottom left* (Ayatollahi et al. 2011c)

4.5.1 Effects of Filler Loading

For all polymer/nano-particle systems, there will be an optimum amount of the nanoparticles beyond which there will be a reduction in the toughness as the stiffness and strength increase. It is possible to tailor the properties of polymer composite with appropriate fillers. And it is reasonable to expect that different volume content of nanoparticles added into polymers may exert different influence on the final material's friction and wear properties (Wang and Pei 2008).

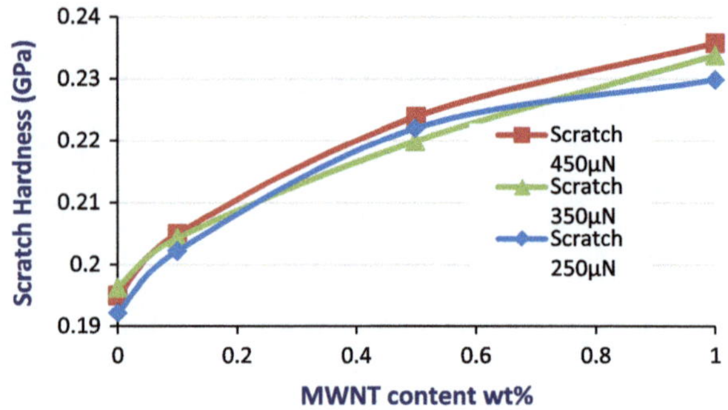

Fig. 4.23 Scratch hardness values obtained for pure epoxy and for nanocomposites (Ayatollahi et al. 2012a)

Ayatollahi et al. investigated the effects of multi-walled carbon nanotubes content on the properties of epoxy-based nanocomposites using the nanoindentation and nanoscratch techniques. In this study, the nanoindentation and nanoscratch tests were conducted on MWCNT reinforced epoxy nanocomposites with three different percent contents of MWCNTs in order to determine their near surface mechanical properties such as elastic modulus, hardness, friction, and self-healing properties. Results showed that in higher nanofiller contents, the plasticity index values, the amount of pile-ups and the scratch depths decreased while the elastic recovery increased. The frictional properties of nanocomposites showed that the ploughing term of friction solely depends on the elastic-plastic behavior. An equation was also proposed to correlate the ploughing friction with the plasticity index (Ayatollahi et al. 2012a). Scratch hardness values of pure epoxy and three nanocomposites in this research are illustrated in Fig. 4.23.

According to Fig. 4.23, while increasing the nanofillers leads to improvement in the values of scratch hardness consistently, the rate of increase is subsided (i.e. the slope of the curve reduces as the amount of MWCNT increases). A similar trend was observed in the nano-indentation analysis, which can be attributed to the presence of more agglomeration in the nanocomposites filled with higher contents of MWCNT. Several parameters are involved in this rising trend. As the indenter slides in the nanocomposites, the nanotubes may experience axial compression, bending or buckling depending on their direction. The energy required to overcome these obstacles increase the scratch hardness of the nanocomposites in higher contents of the nanofillers. In addition to the superior elastic properties of MWCNTs in the epoxy, tribochemical reactions between fillers and slider forms a thin and stable transfer film on the counterface. Formation of this thin film in the interface of indenter and nanocomposite protects the soft polymer from aggressive damage induced by the hard indenter. However, incorporation of higher contents of MWCNTs leads to a number of loosened particles in the tribosystem. In fact, presence of these aggregations weaken the main mechanisms involving in the load transfer from fillers such as micromechanical interlocking, chemical bonding between filler and matrix and even the van der Waals bondings (Ayatollahi et al. 2012a).

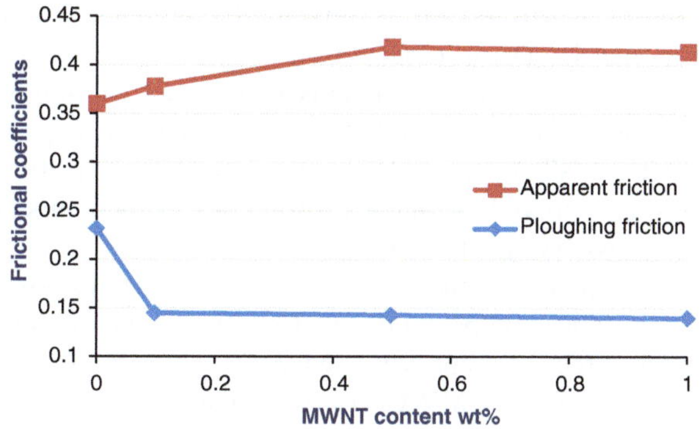

Fig. 4.24 Apparent and ploughing coefficients of friction in terms of MWCNT percent content (Ayatollahi et al. 2012a)

Figure 4.24 shows the values of apparent, ploughing and the interfacial coefficients of friction in terms of the MWCNT content. It is seen that the apparent friction increases with the addition of the nanofillers. Meanwhile, Fig. 4.24 also shows that the apparent ploughing friction is decreasing for higher MWCNTs contents (Ayatollahi et al. 2012a).

Cai et al. in their investigation on polyimide (PI)/singlewalled CNT nanocomposites have shown that they exhibit lower friction coefficients and wear volume losses than neat polyimide under dry sliding conditions against steel at a sliding velocity of 0.43 m/s, test duration 1.5 h, and normal loads in the range of 50–290 N (Cai et al. 2004). The differences were attributed to the quality of transfer films (Fig. 4.25) in neat polyamide and nanocomposites, which led to adhesive wear damage in neat polyimide but significantly reduced scuffing and adhesion in the nanocomposites. However, the reasons for these differences in transfer films and the roles of CNT on the wear processes are unclear. Also, the wear volume losses are less compared to the neat material even at higher CNT volume fractions. However, it should be noted that with more than 1 wt% CNT, it is hard to

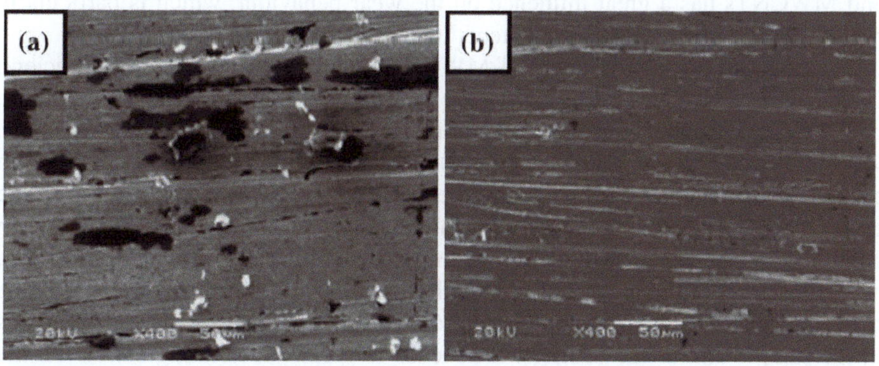

Fig. 4.25 Sample SEM pictures of transfer films generated on the steel ring surface in **a** neat polyimide and **b** polyimide/CNT nanocomposite with 8 wt% of CNT (sliding conditions: load 200 N; velocity 0.431 m/s; test duration 90 min) (Cai et al. 2004)

control the dispersion or separate the aggregated and entangled tubes in the polymer matrix (Dasari et al. 2009).

Wang et al. showed results of reduced friction and improved wear resistance in polyacrylonitrile-methylmethacrylate/CNT copolymer nanocomposites (up to 3 wt% CNTs) prepared via an in situ polymerization process and attributed them to the improved mechanical properties of the nanocomposites, particularly microhardness (Wang et al. 2008).

Campo et al. studied the wear behaviour of epoxy matrix composites with different types and percentages of multiwalled carbon nanotubes (MWCNTs). In this work three different types (NC3100, NC3150 and NC3152) and percentages (0.1 and 0.5 wt%) of MWCNTs were dispersed into an epoxy resin by a calendering process. The tribological properties of epoxy-MWCNTs nanocomposites were investigated using "pin-on disc" wear testing machine under different conditions. Compared with neat epoxy, the composites with MWCNTs showed a lower mass loss, friction coefficient and wear rate, and these parameters decreased with the increase of MWCNT percentage. Also, the results demonstrated that the epoxy composites with 0.5 wt% of amino-MWCNT (NC3152) have the best tribological properties (Campo et al. 2015).

The influence of MWCNTs content on the mass loss, wear rate and friction coefficient is shown in Fig. 4.26.

It is clear, for both types of MWCNTs that the mass loss and wear rate decrease with respect to the neat epoxy with the increase of MWCNTs percentage. This reduction is more pronounced with only 0.1 wt% of MWCNTs and gradually decreases for percentage between 0.1 and 0.5 wt%. The properties of carbon nanotubes give to the resin epoxy an improvement in the wear behaviour due to being extremely hard in the direction of the axis, but at the same time flexible, and also to its lubricating effect. Comparing two types of carbon nanotubes, we can observe that for the same percentage of MWCNTs, the values are slightly lower in the case of functionalised MWCNT (NC3152). This difference is greater for higher percentages (0.5 wt%). The result indicates that stronger interaction between matrix and MWCNTs has a great influence on the wear behaviour, which is particularly noticeable at higher MWCNTs contents (Campo et al. 2015).

Figure 4.26c shows the effect of MWCNTs content on the friction coefficient. It can be observed that at low MWCNTs percentage, it remained almost constant, being slightly higher for materials containing NC3152 MWCNTs. However, when the percentage is higher (0.5 wt%) a sharp drop in the value occurs in both cases. This behaviour could be due to the tribological layer (debris) formed by the epoxy resin detached and the carbon nanotubes torn during the test. These carbon nanotubes are present in the debris having a lubricating effect which produces a reduction in the friction coefficient. The low coefficient of friction observed in the materials with a 0.5 wt% could be due to the higher amount of carbon nanotubes in the sample and in the debris, which leads to greater lubricating effect. On the other hand, the increase of the friction coefficient that occurs in the materials with 0.1, 0.2 and 0.3 wt% in relation to the epoxy matrix may be because debris with MWCNTs can act as a third body in the wear mechanism, thus further increasing

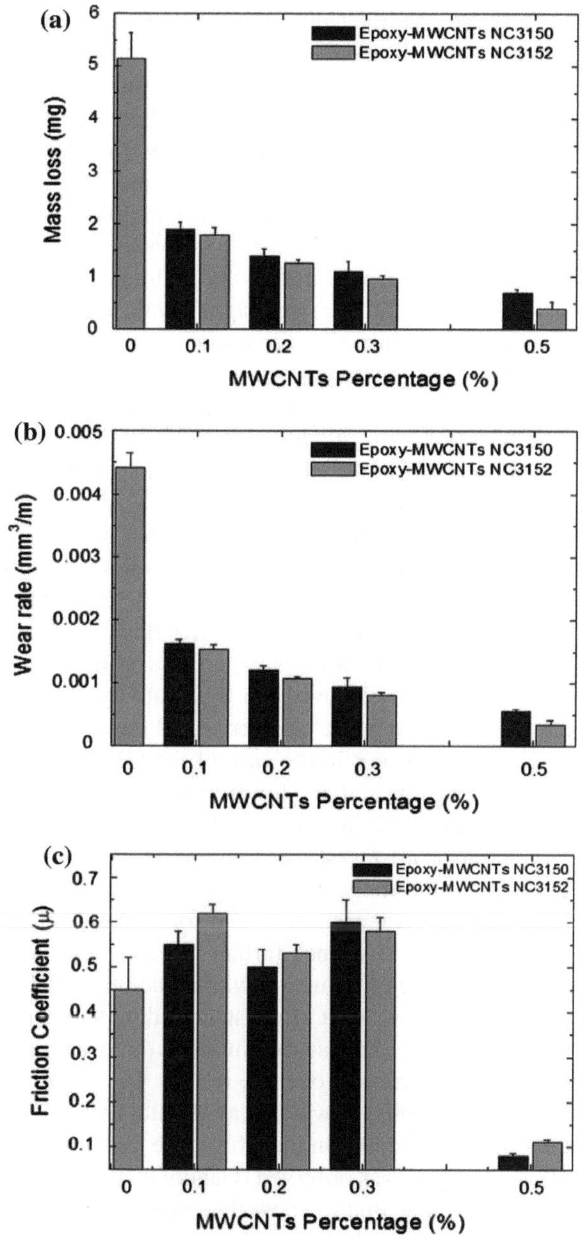

Fig. 4.26 The values of mass loss (**a**), wear rate (**b**) and friction coefficient (**c**) of epoxy-MWCNTs nanocomposites with different percent contents of MWCNTs (Campo et al. 2015)

real contact surface, and producing a slight increase in the friction coefficient (Dasari et al. 2009). This effect is diminished as the amount of MWCNTs increases, where the property as a lubricant effect of MWCNTs is higher (Campo et al. 2015).

Cui et al. investigated the friction and wear behaviour of epoxy-MWCNTs nanocomposites at different sliding speeds under different applied loads (40–120 N). However, they investigated lower percentages (0–0.5 wt%) and the influence of the functionalisation of carbon nanotubes with amino and carboxyl group. They showed that compared with neat epoxy, the nanocomposites with MWCNTs had a lower friction coefficient and wear rate, and the wear rate decreased with the increase of MWCNTs loading (Cui et al. 2013). Dong et al. studied the influence of different contents of MWCNTs (0–4 wt%) on tribological properties of the epoxy-MWCNTs nanocomposites in dry sliding against plain carbon steel. They found that the MWCNTs increased significantly wear resistance and reduced coefficient of friction of the nanocomposites, concluding that the nanocomposite with 1.5 wt% MWCNTs showed the best wear behaviour (Dong et al. 2005).

Kalin et al. studied Wear and friction behaviour of poly-ether-ether-ketone (PEEK) filled with graphene, WS2 and CNT nanoparticles. Different nanoparticles were added to the PEEK matrix: WS_2 fullerene-like (WS_2F), WS_2 needle-like (WS_2N), carbon nanotubes (CNT) and grapheme nanopowder (GNP). The results of dry-sliding tribological tests show that the material and the morphology of the nanoparticles have a crucial effect on both the wear and the coefficient of friction, primarily by affecting their macroscopic hardness, as well as the thickness and the surface coverage of their transfer films. The WS_2-based nanoparticles outperformed the carbon-based nanoparticles in terms of wear performance; compared to PEEK, WS_2F improved the wear rate by 10 % and the WS_2N by 60 %, while the carbon-based particles deteriorated the wear behaviour by 20 % (CNT) and as much as three times in the case of the GNP. The addition of WS_2F nanoparticles and the CNT resulted in a lower coefficient of friction compared to the WS_2N particles and the GNP (Kalin et al. 2014).

Gandhi et al. investigated the addition of CNT in PP in wear performance. The results normally indicated that the increase of carbon nanotube percentage increases the wear resistance and for finding the optimal percentage content with respect to wear resistance further research is required. The specific wear rate and their percentage of reduction with an addition of CNT were recorded. The experiments were carried out at constant sliding speed of 1 m/s, the loads ranged from 10 to 50 N. In each case minimum of three samples were tested and the observed values are the average of these results. The wear rate of PP/CNT-93/07 composite at 50 N was 0.0029 g which was about 44 % lower compared with unfilled PP for the same conditions. On the whole, the higher the CNT proportions the greater the wear resistance was observed (Gandhi et al. 2013).

Experimental studies show that nanodiamond (ND) is more effective than CNT to improve tribological properties of nanocomposites. For example, as reported by Ayatollahi et al., the addition of only 0.1 wt% of ND decreases the coefficient of friction and the wear rate 50 and 84 %, respectively (Ayatollahi et al. 2012b). From the above studies on polymer/CNT nanocomposites, it is realized that the mechanical and tribological properties can be enhanced, but the understanding of the mechanisms of wear and friction are still at the early stage. Significant contributions are needed to improve the fundamental tribological knowledge of these novel functional materials.

4.5.2 Effects of CNT Aspect Ratio

Ayatollahi et al. investigated the effects of multi-walled carbon nanotubes content and aspect ratio on the properties of epoxy-based nanocomposites. Four types of MWCNTs with different diameters of 20 < d < 30, 30 < d < 40, 40 < d < 50 and 50 < d < 60 nm were used. In addition, the intrinsic length of all MWCNTs was 10 < l < 30 μm with the average length of 25 μm, as reported by the supplier (Nanostructured and Amorphous Materials Inc.). Instrumented indentation tests were performed with the penetration load of 500 μN and the constant loading rate of 15 μN/s, using the testing instrument, (Triboscope system Hysitron Inc., USA) according to the procedure described in ISO 14577 standard. In order to explore the effect of interfacial characteristics on the nanocomposites properties, a comparison between the results obtained from the nanocomposites containing 0.5 wt% MWCNTs of different aspect ratios was conducted. Results showed that increase in both the content and the aspect ratio of MWCNTs improved the elastic modulus as well as the normal and scratch hardness of the epoxy-based nanocomposites (Ayatollahi et al. 2012a).

Liu et al. used three varieties of multi-walled CNTs (two of which were claimed to have different diameters, 20–30 nm versus 30–50 nm and the other is a carboxylatedmultiwalled CNT) in BMI and characterized their wear properties. Although they concluded that tribological properties of these nanocomposites depended on the diameter and functionalization of CNTs, the results/evidence shown are not at all conclusive. They also noted a change in wear mechanism with the addition of CNTs from neat material and claimed this to be a result of the dispersion and improved interfacial strength between filler and matrix. Also results showed that the addition of MWCNTs in BMI resin decreases the friction coefficient of the resin no matter what kind of MWCNTs is added (Liu et al. 2007a, b, c).

Lee et al. studied The friction and wear of polyamide 6,6 (PA66) nanocomposites containing 1 wt% of carbon nanotubes (CNTs) with different lengths using a block-on-ring tribometer at different sliding velocities. In the sliding tests performed at temperature below 110 °C the coefficient of friction (COF) tended to be decreased by CNTs addition, while it was slightly increased as the CNT length increased. However, at temperatures beyond 110 °C, the COF increased with the addition of CNTs and increased, even more, with longer CNTs. Wear resistance was also improved by long CNTs, and this improvement was particularly evident at elevated temperatures (Lee et al. 2014).

It is clear that for a constant content of nanotubes, as the tube diameter decreases, the number of nanoparticles grows; therefore, the interfacial area of the nanocomposite system increases. This provides more chance for stress transmission between the matrix environment and the nanofillers. Meanwhile, previous studies have shown that a stronger interface would be established between the matrix and the nanofillers of smaller diameters in the polymeric nanocomposites (Manoharan et al. 2009). Larger and more efficient interfacial areas lead to better load transfer, and consequently better tribology properties of nanocomposites.

4.5.3 Effects of Secondary Nanoparticles

Shen et al. conducted an investigation on the effect of the hybrid of multi-wall carbon nanotubes (MWCNTs) and graphene oxide (GO) nanosheets on the tribological performance of epoxy composites at low GO weight fractions of 0.05–0.5 phr. The MWCNT amount is kept constant at 0.5 phr, which is typical for CNT/epoxy composites with enhanced mechanical properties. Friction and wear tests against smooth steel show that the introduction of 0.5 phr MWCNTs into the epoxy matrix increases the friction coefficient and decreases the specific wear rate. When testing the tribological performance of MWCNT/GO hybrids, it is shown that at a high GO amount of 0.5 phr, the friction coefficient is decreased below that of the neat matrix whereas the wear rate is increased above that of the neat matrix. At an optimal hybrid formulation, i.e., 0.5 phr MWCNTs and 0.1 phr GO, a further increase in the friction coefficient and a further reduction in the specific wear rate are observed. The specific wear rate is reduced by about 40 % down to a factor of 11 relative to the neat epoxy when the GO content is 0.1 phr (Shen et al. 2014). Other studies focus on tribological properties of other nanoparticles hybrid nanocomposites. More research studies are needed to improve the fundamental tribological knowledge of these hybrid materials.

4.6 Electrical Properties

A number of studies have been done related to the electrical properties of CNT/polymer nanocomposites which show that combination of CNTs and polymer matrix offers new attractive electrical properties and conducting polymers. These polymers can be used for various engineering applications like, electrical conducting adhesives, antistatic coating and films, electromagnetic interference shielding materials for electronic devices, thermal interface materials, etc. These conducting nanocomposites follow the principle of percolation theory which basically explains the transition from insulator to conductor in materials. Accordingly, a small increase in loading will result in a drastic increase in conductivity. After a critical amount of filler which is known as electrical conductivity of the composites jumps up several orders of magnitude (Mittal et al. 2015). This happens due to the formation of continuous conducting networks. Above this percolation transition range, conductivity of the composites get saturated due to the multiple electrons networks and below this range, electrons paths do not exist (Ma et al. 2010).

Due to very high aspect ratio and splendid electrical properties CNTs avert these problems. Sandler et al. reported the ultra-low electrical percolation threshold i.e. 0.0025 wt% in aligned CNT/epoxy composites (Sandler et al. 2003). Percolation threshold depends upon various factors such as aspect ratio, dispersion, alignment, functionalization, processing, polymer type, etc. (Mittal et al. 2015). For example, Allaoui et al. showed that percolation threshold value for

MWCNT/Epoxy composites is around 0.5–1 wt% (Allaoui et al. 2002). Aguilar et al. studied the clustering effect of CNTs on electrical conductivity of polysulfone films loaded with 0.05–0.75 wt%. Percolation threshold for homogenously dispersed and agglomerated films was reported to be 0.11 and 0.068 % (w/w), respectively. The conductivity of the agglomerated films was found 2–4 orders of magnitudes higher than the uniformly dispersed films which can be explained by increased density CNT–CNT junctions (Aguilar et al. 2010). He et al. studied temperature dependent electrical properties of MWCNT/high density polyethylene (HDPE) composites prepared by solution-precipitation. The PTC intensity for MWCNT/HDPE composites can reach 1.1×10^4 by loading 5.4 wt% MWCNTs (He et al. 2005). However, some studies indicate that electrical conductivity of matrix significantly decreases by chemical functionalization of CNTs due to imbalance in the polarization effect and physical structure defects (Sulong et al. 2009). Nanocomposites based on PMMA and MWCNTs as a filler show a significant enhancement in the electrical conductivity (Spitalsky et al. 2010). Kim et al. reported a value of ≈3000 S/mat 0.4 wt% for extremely low percolation threshold (≈0.003 wt%) prepared by the solution mixing method (Kim et al. 2004a, b). Furthermore, the addition of MWCNTs to the polyester matrix PC by the melt mixing technique induces a dramatic increase of about 16 orders of magnitude in the maximum conductivity; reaching the value of ≈1000 S/m at 15 wt% (Pötschke et al. 2004).

Safdari et al. studied electrical and thermal transport properties of hybrid polymeric nanocomposites based on carbon nanotubes and graphite nanoplatelets. A number of CNT/polymer, GNP/polymer and hybrid CNT/GNP/polymer nanocomposites specimens were processed. For the CNT/polymer and the GNP/polymer specimens, five different configurations were prepared comprising 1.0–5.0 %wt of the CNTs and the GNPs, respectively. For the hybrid CNT/GNP polymer specimens five different configurations were selected as well; each comprising (X)%wt GNP and (X/10)%wt CNT were X = 1.0–5.0. In this study, the average electrical conductivities of the specimens were measured via the four-point probe method as reported in Fig. 4.27. The electrical conductivity for GNP/epoxy samples with 1.0 3.0 %wt GNP are not reported in this figure since the minimum possible reading by the utilized equipment was 10^{-7} S/m which was higher than the conductivity of these specimens. In the current study, the electrical conductivity equivalent to 10^{-7} S/m is also selected as the percolation threshold criterion which corresponds to six orders of magnitudes enhancement in the electrical conductivity over that for the base epoxy host matrix (Safdari and Al-Haik 2013).

It can be concluded that experimental measurements suggest that by adding minute amounts of the CNTs (10 % compared to GNPs) as the auxiliary conductive phase, the percolation threshold diminishes at least by 50 % and the electrical conductivity increases by several orders of magnitudes. This substantial improvement can be attributed primarily to the formation of the highly effective hybrid CNT/GNP networks (Safdari and Al-Haik 2013).

Li et al. studied the effects of hybrid carbon nanotubes (CNT) and graphite nanoplatelets (GNPs) on the electrical properties of epoxy nanocomposites. When

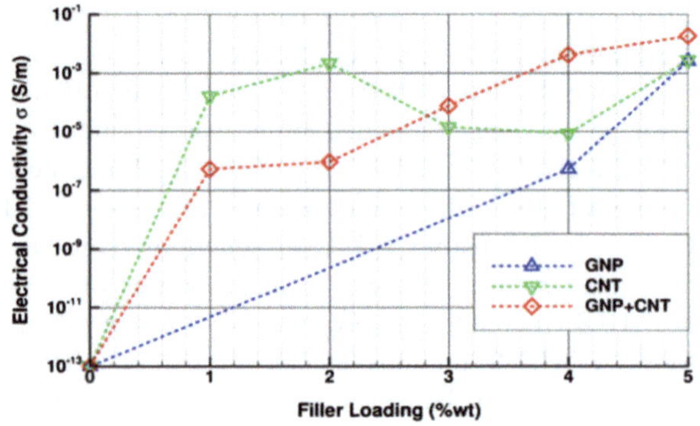

Fig. 4.27 The values of electrical conductivity obtained for CNT/epoxy, GNP/epoxy and CNT/GNP/epoxy specimen measured experimentally by using the four-point probe technique (Safdari and Al-Haikr 2013)

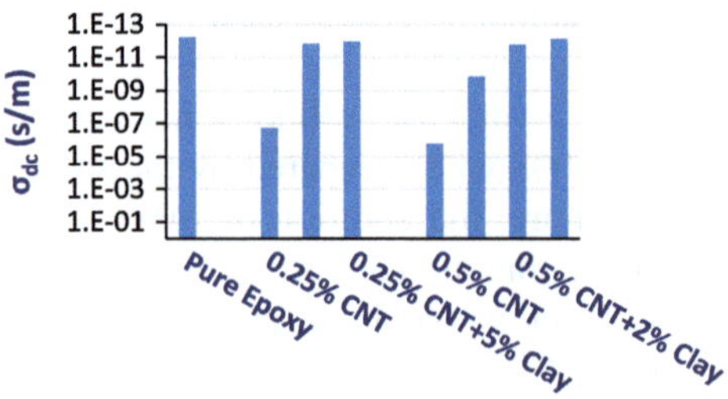

Fig. 4.28 The values of electrical conductivity obtained for pure epoxy and MWCNT-nanoclay/epoxy (Ayatollahi et al. 2011e)

the total reinforcement contents is fixed at 2 wt%, the nanocomposite containing 1 % GNP and 1 % CNT achieves the highest electrical conductivity of 4.7×10^{-3} S/cm, which is more than two orders of magnitude higher than that of nanocomposites with 2 wt% GNP alone (Li et al. 2008).

Ayatollahi et al. studied the effect of simultaneous presence of nanoclay and MWCNT on the electrical prosperities of the resulting nanocomposite. In this study in order to evaluate the effect of presence of different particles on the electrical properties of resulted nanocomposite, pure epoxy and seven different types of nanocomposites were prepared for the electrical tests. Figure 4.28 shows the results of electrical test related to the epoxy/MWCNT-nanoclay nanocomposites at various filler contents. It is seen that by addition of 0.25 wt% MWCNT, the electrical conductivity dramatically increases which means that percolation threshold is already obtained. That is to say, conductive network is built in the matrix and electrons can be conducted through it and as a result the electrical resistance is decreased noticeably (Ayatollahi et al. 2011e).

On the other hand, addition of nanoclay to the MWCNT/epoxy composite generally causes a considerable reduction in the electrical conductivity. For nanocomposite containing 0.25 wt% MWCNT by addition of the nanoclay particles the conductivity of the nanocomposite completely disappeared. For nanocomposites containing 0.5 wt% MWCNT, by addition of 1 wt% nanoclay the conductivity drops 10^4 times. By further addition of nanoclay the conductivity vanished altogether (Ayatollahi et al. 2011e).

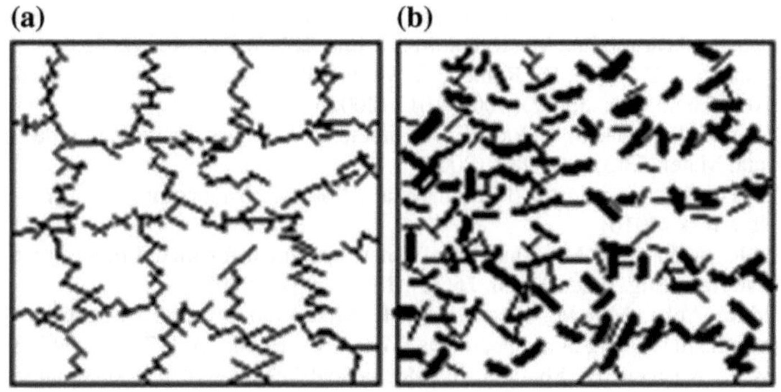

Fig. 4.29 Conductive network shown schematically for **a** MWCNT/epoxy nanocomposite **b** MWCNT-nanoclay/epoxy (Ayatollahi et al. 2011e)

Figure 4.29 schematically illustrates the conductive network in epoxy/MWCNT and epoxy/MWCNT-nanoclay nanocomposites. As can be seen in Fig. 4.29b, the nanoclay particles are dispersed in the matrix and block the contact paths between MWCNTs and consequently MWCNTs are not able to make a fully conductive network. Therefore, the electrons have to tunnel between the MWCNT particles which cause a huge reduction in the electrical conductivity. By adding more than 1 wt% nanoclay in the nanocomposite, the ability of electrons to tunnel completely disappears (Ayatollahi et al. 2011e). While some researchers observed the same trend (Kim et al. 2010), this finding is dissimilar to what is reported in some other studies (Liu and Grunlan 2007; Santangelo et al. 2011).

References

Aguilar, J.O., Bautista-Quijano, J.R., Avilés, F.: Influence of carbon nanotube clustering on the electrical conductivity of polymer composite films. Polym. Lett. **4**(5), 292–299 (2010)

Aihara, J.: Lack of superaromaticity in carbon nanotubes. Phys. Chem. **98**(39), 9773–9776 (1994)

Ajayan, P.M., Schadler, L.S., Giannaris, G., Rubio, A.: Single-walled carbon nanotube–polymer composites: strength and weakness. Adv. Mater. **12**(10), 750–753 (2000)

Aliha, M.R.M., Ayatollahi, M.R.: On mixed-mode I/II crack growth in dental resin materials. Scripta Mater. **59**(2), 258–261 (2008)

Aliha, M.R.M., Ayatollahi, M.R.: Geometry effects on fracture behaviour of polymethyl methacrylate. Mater. Sci. Eng. A **527**(3), 526–530 (2010)

Alishahi, E., Shadlou, S., Doagou-R, S., Ayatollahi, M.R.: Effects of carbon nanoreinforcements of different shapes on the mechanical properties of epoxy-based nanocomposites. Macromolecular **298**(6), 670–678 (2013)

Allaoui, A., Bai, S., Cheng, H.M., Bai, J.B.: Mechanical and electrical properties of a MWNT/epoxy composite. Compos. Sci. Technol. **62**(15), 1993–1998 (2002)

Al-Saleh, M.H., Saadeh, W.: Hybrids of conductive polymer nanocomposites. Mater. Des. **52**, 1071–1076 (2013)

Andrews, R., Jacques, D., Minot, M., Rantell, T.: Fabrication of carbon multiwall nanotube/polymer composites by shear mixing. Macromol. Mater. Eng. **287**(6), 395–403 (2002a)

Andrews, R., Jacques, D., Qian, D., Rantell, T.: Multiwall carbon nanotubes: synthesis and application. Acc. Chem. Res. **35**(12), 1008–1017 (2002b)

Araki, W., Nemoto, K., Adachi, T., Yamaji, A.: Fracture toughness for mixed mode I/II of epoxy resin. Acta. Mater. **53**(3), 869–861 (2005)

Ashton, H.C.: The Incorporation of Nanomaterials. Polymer Nanocomposites, pp. 21–44. CRC Press, New York (2010)

Ayatollahi, M.R., Aliha, M.R.M.: Analysis of a new specimen for mixed mode fracture tests on brittle materials. Eng. Fract. Mech. **76**(11), 1563–1573 (2009)

Ayatollahi, M.R., Aliha, M.R.M., Hassani, M.M.: Mixed mode brittle fracture in PMMA—an experimental study using SCB specimens. Mater. Sci. Engng. A **417**(1–2), 348–356 (2006)

Ayatollahi, M.R., Alishahi, E., Shadlou, S.: Mechanical behavior of nanodiamond/epoxy nanocomposites. Int. J. Fract. **170**(1), 95–100 (2011a)

Ayatollahi, M.R., Shadlou, S., Shokrieh, M.M.: Multiscale modeling for mechanical properties of carbon nanotube reinforced nanocomposites subjected to different types of loading. Compos. Struct. **93**(9), 2250–2259 (2011b)

Ayatollahi, M.R., Shadlou, S., Shokrieh, M.M.: Fracture toughness of epoxy/multi-walled carbon nanotube nano-composites under bending and shear loading conditions. Mater. Des. **32**(4), 2115–2124 (2011c)

Ayatollahi, M.R., Shadlou, S., Shokrieh, M.M., Chitsazzadeh, M.: d. Effect of multi-walled carbon nanotube aspect ratio on mechanical and electrical properties of epoxy-based nanocomposites. Polym. Test. **30**(5), 548–556 (2011d)

Ayatollahi, M.R., et al.: Mechanical and electrical properties of epoxy/MWNT-nanoclay nanocomposites. Iran. Polym. J. **20**(10), 832–843 (2011e)

Ayatollahi, M.R., Shadlou, S., Shokrieh, M.M.: Mixed mode brittle fracture in epoxy/multi-walled carbon nanotube nanocomposites. Eng. Fract. Mech. **78**(14), 2620–2632 (2011f)

Ayatollahi, M.R., Shadloua, S., Shokrieh, M.M.: Correlation between aspect ratio of MWCNTs and mixed mode fracture of epoxy based nanocomposites. Mater. Sci. Eng. A **528**(19–20), 6173–6178 (2011g)

Ayatollahi, M.R., Doagou-Rad, S., Shadlou, S.: Nano-/microscale investigation of tribological and mechanical properties of epoxy/MWNT nanocomposites. Macromol. Mater. Eng. (Wiley) **297**(7), 689–701 (2012a)

Ayatollahi, M.R., Alishahi, E., Doagou-R, S., Shadlou, S.: Tribological and mechanical properties of low content nanodiamond/epoxy nanocomposites. Compos. B **43**(8), 3425–3430 (2012b)

Bahr, J.L., et al.: Functionalization of carbon nanotubes by electrochemical reduction of aryl diazonium salts: a bucky paper electrode. J. Am. Chem. Soc. **123**(27), 6536–6542 (2001)

Bai, J.: Evidence of the reinforcement role of chemical vapour deposition multi-walled carbon nanotubes in a polymer matrix. Carbon **41**(6), 1325–1328 (2003)

Barber, A.H., Cohen, S.R., Kenig, S., Wagner, H.D.: Interfacial fracture energy measurements for multi-walled carbon nanotubes pulled from a polymer matrix. Compos. Sci. Technol. **64**(15), 2283–2289 (2004)

Bhattacharjee, D., Knott, J.F.: Effect of mixed mode I and II loading on the fracture surface of polymethyl methacrylate (PMMA). Int. J. Fract. **72**(4), 359–381 (1995)

Bhattacharyya, S., Sinturel, C., Salvetat, J.P., Saboungi, M.L.: Proteinfunctionalized CNT polymer composites. Appl. Phys. Lett. **86**, 113104 (2005)

Bhattacharyya, S., Salvetat, J.P., Saboungi, M.L.: Reinforcement of semicrystalline polymers with collagen modified SWCNTs. Appl. Phys. Lett. **88**, 233119 (2006)

Bhuiyan, M.A., Pucha, R.V., Karevan, M., Kalaitzidou, K.: Tensile modulus of carbon nanotube/polypropylene composites—a computational study based on experimental characterization. Comput. Mater. Sci. **60**(8), 2347–2353 (2011)

Bhuiyan, M.A., et al.: Defining the lower and upper limit of the effective modulus of CNT/polypropylene composites through integration of modeling and experiments. Compos. Struct. **95**, 80–87 (2013)

Bin, Y., Kitanaka, M., Zhu, D., Matsuo, M.: Development of highly oriented polyethylene filled with aligned carbon nanotubes by gelation/crystallization from solutions. Macromolecules **36**(6), 6213–6219 (2003)

Blake, R., et al.: Reinforcement of poly(vinyl chloride) and polystyrene using chlorinated polypropylene grafted carbon nanotubes. J. Mater. Chem. **16**(43), 4206–4213 (2006)

Blanco, J., García, E.J., Villoria, R.G., Wardle, B.L.: Limiting mechanisms of mode I interlaminar toughening of composites reinforced with aligned carbon nanotubes. J. Compos. Mater. **43**(8), 825–841 (2009)

Blond, D., et al.: Enhancement of modulus, strength, and toughness in poly(methyl methacrylate)-based composites by the incorporation of poly(methyl methacrylate)-functionalised nanotubes. Adv. Funct. Mater. **16**(12), 1608–1614 (2006)

Böger, L., Sumfleth, J., Hedemann, H., Schulte, K.: Improvement of fatigue life by incorporation of nanoparticles in glass fibre reinforced epoxy. Compos. A **41**(10), 1419–1424 (2010)

Breton, Y., et al.: Mechanical properties of multiwall carbon nanotubes/epoxy composites: influence of network morphology. Carbon **42**(5–6), 1027–1030 (2004)

Brian, J., Sinha, B.K.: Tribological applications of polymers and their composites: past, present and future prospects. Tribology of Polymeric Nanocomposites, pp. 7–11. Elsevier Science, s.l. (2008)

Cadek, M.: Mechanical and thermal properties of multiwalled carbon nanotube reinforced polymer composites. San Diego, s.n. (2002)

Cadek, M., et al.: Morphological and mechanical properties of carbon-nanotube-reinforced semicrystalline and amorphous polymer composites. Appl. Phys. Lett. **81**(27), 5123–5125 (2002)

Cadek, M., et al.: Reinforcement of polymers with carbon nanotubes: the role of nanotube surface area. Nano Lett. **4**(2), 353–356 (2004)

Cai, H., Yan, F.Y., Xue, Q.J.: Investigation of tribological properties of polyimide/carbon nanotube nanocomposites. Mater. Sci. Eng. A **364**(1–2), 94 (2004)

Campo, M., Jiménez-Suárez, A., Ureña, A.: Effect of type, percentage and dispersion method of multi-walled carbon nanotubes on tribological properties of epoxy composites. Wear **324**, 100–108 (2015)

Chang, T.E., et al.: Microscopic mechanism of reinforcement in SWCNT–polypropylene nanocomposite. Polymer **46**(2), 439–444 (2005)

Chen, J., et al.: Solution properties of single-walled carbon nanotubes. Science **282**(5386), 95–98 (1998)

Chen, L., Pang, X.J., Yu, Z.L.: Study on polycarbonate/multi-walled carbon nanotubes composite produced by melt processing. Mater. Sci. Eng. A **457**(1–2), 287–291 (2007)

Chen, Z., et al.: Improving the mechanical properties of multiwalled carbon nanotube/epoxy nanocomposites using polymerization in a stirring plasma system. Compos. Part A **56**, 172–180 (2014)

Coleman, J.N., et al.: High performance nanotube-reinforced plastics: understanding the mechanism of strength increase. Adv. Funct. Mater. **14**(8), 791–798 (2004)

Coleman, J.N., Khan, U., Blau, W.J., Gun'ko, Y.K.: Small but strong: a review of the mechanical properties of carbon nanotube–polymer composites. Carbon **44**(9), 1624–1652 (2006)

Cooper, C.A., Young, R.J., Halsall, M.: Investigation into the deformation of carbon nanotubes and their composites through the use of Raman spectroscopy. Compos. A **32**(3–4), 401–411 (2001)

Cooper, C.A., et al.: Distribution and alignment of carbon nanotubes and nanofibrils in a polymer matrix. Compos. Sci. Technol. **62**(7–8), 1105–1112 (2002)

Cui, L.J., et al.: Functionalization of multi-wall carbon nanotubes to reduce the coefficient of the friction and improve the wear resistance of multi-wall carbon nanotube/epoxy composites. Carbon **54**, 277–282 (2013)

Dai, L., Mau, A.W.H.: Controlled synthesis and modification of carbon nanotubes and C60: carbon nanostructures for advanced polymeric composite materials. Adv. Mater. **13**(12–13), 899–913 (2001)

Dalton, A.B., et al.: Super-tough carbon-nanotube fibres. Nature **423**, 703 (2003)
Dasari, A., Yu, Z.Z., Mai, Y.W.: Fundamental aspects and recent progress on wear/scratch damage in polymer nanocomposites. Mater. Sci. Eng. R Rep. **63**(2), 31–80 (2009)
Dondero, W.E., Gorga, R.E.: Morphological and mechanical properties of CNT–polymer composites via melt compounding. J. Polym. Sci. Part B: Polym. Phys. **44**(5), 864–878 (2006)
Dong, B., Yang, Z., Huang, Y., Li, H.L.: Study on tribological properties of multiwalled carbon nanotubes/epoxy resin nanocomposite. Tribol. Lett. **20**(3–4), 251–254 (2005)
Du, F., Fischer, J.E., Winey, K.I.: Coagulation method for preparing singlewalled carbon nanotube/poly(methyl methacrylate) composites and their modulus, electrical conductivity, and thermal stability. J. Polym. Sci. Part B: Polym. Phys. **41**(24), 3333–3338 (2003)
Faber, K.T., Evans, A.G.: Crack deflection processes—I. Theory Acta Metall. **31**(4), 565–576 (1983)
Fiedler, B., et al.: Fundamental aspects of nano-reinforced composites. Compos. Sci. Technol. **66**(16), 3115–3125 (2006)
Fritzsche, J., Lorenz, H., Klüppel, M.: CNT based elastomer-hybrid-nanocomposites with promising mechanical and electrical properties. Macromol. Mater. Eng. **294**(9), 551–560 (2009)
Gandhi, R.A., Palanikumar, K., Ragunath, B.K., Davim, J.P.: Role of carbon nanotubes (CNTs) in improving wear properties of polypropylene (PP) in dry sliding condition. Mater. Des. **48**, 52–57 (2013)
Gojny, F.H., Nastalczyk, J., Roslaniec, Z., Schulte, K.: Surface modified multi-walled carbon nanotubes in CNT/epoxy-composites. Chem. Phys. Lett. **370**(5–6), 820–824 (2003)
Gojny, F.H., et al.: Carbon nanotube-reinforced epoxy-composites: enhanced stiffness and fracture toughness at low nanotube content. Compos. Sci. Technol. **64**(15), 2663–2671 (2004)
Gojny, F.H., Wichmann, M.H.G., Fiedler, B., Schulte, K.: Influence of different carbon nanotubes on the mechanical properties of epoxy matrix composites—a comparative study. Compos. Sci. Technol. **65**(15–16), 2300–2313 (2005)
Gorga, R.E., Cohen, R.E.: Toughness enhancements in poly(methyl methacrylate) by addition of oriented multiwall carbon nanotubes. J. Polym. Sci. Part B: Polym. Phys. **40**(14), 2690–2702 (2004)
Gorrasi, J., et al.: Incorporation of carbon nanotubes into polyethylene by high energy ball milling: morphology and physical properties. J. Polym. Sci. Part B: Polym. Phys. **45**(5), 597–606 (2007)
Grady, B.P., Pompeo, F., Shambaugh, R.L., Resasco, D.E.: Nucleation of polypropylene crystallization by SWCNTs. J. Phys. Chem. B **106**(23), 5852–5858 (2002)
Guo, P., et al.: Fabrication and mechanical properties of well-dispersed multiwalled carbon nanotubes/epoxy composites. Compos. Sci. Technol. **67**(15–16), 3331–3337 (2007)
Haggenmueller, R., et al.: Aligned single-wall carbon nanotubes in composites by melt processing methods. Chem. Phys. Lett. **330**(3–4), 219–225 (2000)
Haggenmueller, R., Zhou, W., Fischer, J.E., Winey, K.I.: Production and characterization of polymer nanocomposites with highly aligned single-walled carbon nanotubes. J. Nanosci. Nanotechnol. **3**(1–2), 1105–1110 (2003)
Han, J.H., et al.: CNT buckypaper/thermoplastic polyurethane composites with enhanced stiffness, strength and toughness. Compos. Sci. Technol. **103**, 63–71 (2014)
He, X.J., et al.: Positive temperature coefficient effect in multiwalled carbon nanotube/high-density polyethylene composites. Appl. Phys. Lett. **86**(6), 062112 (2005)
Hough, L.A., Islam, M.F., Janmey, P.A., Yodh, A.G.: Viscoelasticity of singlewall carbon nanotube suspensions. Phys. Rev. Lett. **93**(6), 168102 1–4 (2004)
Hou, Y., et al.: Functionalised few-walled carbon nanotubes for mechanical reinforcement of polymeric composites. ACS Nano **3**(5), 1057–1062 (2009)
Huang, G.L.: Efficient load transfer to polymer-grafted MWCNTs in polymer composites. Adv. Funct. Mater. 487–91 (2004)
Hull, D.: An Introduction to Composite Materials. Cambridge University Press, s.l. (1981)
Jen, M.H.R.: Experiments and Simulations. DEStech Publications, s.l. (2012)

Jia, Z., et al.: Study on poly(methyl methacrylate)/carbon nanotube composites. Mater. Sci. Eng. A **271**(1–2), 395–400 (1999)

Jin, Z., Pramoda, K.P., Xu, G., Goh, S.H.: Dynamic mechanical behavior of melt-processed multi-walled carbon nanotube/poly(methyl methacrylate) composites. Chem. Phys. Lett. **337**(1–3), 43–47 (2001)

Jin, Z., Pramoda, K.P., Goh, S.H., Xu, G.: PVDF-assisted melt blending of MWCNT/PMMA composites. Mater. Res. Bull. **37**(2), 271–278 (2002)

Jose, M.V., et al.: Polypropylene/CNT nanocomposite fibers: process–morphology–property relationships. J. Appl. Polym. Sci. **103**(6), 3844–3850 (2007)

Kalin, M., Zalaznik, M., Novak, S.: Wear and friction behaviour of poly-ether-ether-ketone (PEEK) filled with graphene, WS2 and CNT nanoparticles. Wear WEA101203 (2014)

Kanagaraj, S., et al.: Mechanical properties of high density polyethylene/carbon nanotube composites. Compos. Sci. Technol. **67**(15–16), 3071–3077 (2007)

Kearns, J.C., Shambaugh, R.L.: Polypropylene fibers reinforced with carbon nanotubes. J. Appl. Polym. Sci. **86**(6), 2079–2084 (2002)

Khan, U., Coleman, J.N.: The effect of solvent choice on the mechanical properties of carbon nanotube–polymer composites. Compos. Sci. Technol. 3158–3167 (2007)

Kim, K.H., Jo, W.H.: Improvement of tensile properties of poly(methyl methacrylate) by dispersing multi-walled carbon nanotubes functionalized with poly(3-hexylthiophene)-graft-poly(methyl methacrylate). Compos. Sci. Technol. **68**(9), 2120–2124 (2008)

Kim, H.M., et al.: Electrical conductivity and electromagnetic interference shielding of multi-walled carbon nanotube composites containing Fe catalyst. Appl. Phys. Lett. **84**(4), 589–591 (2004a)

Kim, H.M., et al.: Charge transport properties of composites of multiwalled carbon nanotube with metal catalyst and polymer: application to electromagnetic interference shielding. Curr. Appl. Phys. **4**(6), 577–580 (2004b)

Kim, H.S., et al.: Electrical and mechanical properties of poly(L-lactide)/carbon nanotubes/clay nanocomposites. J. Nanosci. Nanotechnol. **10**(5), 3576–3580 (2010)

Lahelin, M., et al.: In situ polymerized nanocomposites: polystyrene/CNT and poly(methyl methacrylate)/CNT composites. Compos. Sci. Technol. **71**(6), 900–907 (2011)

Lee, S.M., Shin, M.W., Jang, H.: Effect of carbon-nanotube length on friction and wear of polyamide 6,6 nanocomposites. Wear 103–110 (2014)

Leelapornpisit, W., et al.: Effect of carbon nanotubes on the crystallization and properties of polypropylene. J. Polym. Sci. Part B **43**(18), 2445–2453 (2005)

Li, Q., Zaiser, M., Koutsos, V.: Carbon nanotube/epoxy resin composites using a block copolymer as a dispersing agent. Phys. Status Solidi (a) **201**(13), 89–91 (2004a)

Li, X., et al.: Nanomechanical characterization of single-walled carbon nanotube reinforced epoxy composites. Nanotechnology **15**(11), 1416–1423 (2004b)

Li, J., Wong, P.S., Kim, J.K.: Hybrid nanocomposites containing carbon nanotubes and graphite nanoplatelets. Mater. Sci. Eng. A **483–484**, 660–663 (2008)

Liao, Y.H., et al.: Investigation of the dispersion process of SWCNTs/SC15 epoxy resin nanocomposites. Mater. Sci. Eng. A **385**(1–2), 175–181 (2004)

Lin, Y., Taylor, S., Huang, W., Sun, Y.: Characterization of fractions from repeated functionalization reactions of carbon nanotubes. Phys. Chem. **107**(4), 914–919 (2003)

Liu, L., Grunlan, J.C.: Clay assisted dispersion of carbon nanotubes in conductive epoxy nanocomposites. Adv. Funct. Mater. **17**(14), 2343–2348 (2007)

Liu, T., et al.: Morphology and mechanical properties of multiwalled carbon nanotubes reinforced nylon-6 composites. Macromolecules **37**(9), 7214–7222 (2004)

Liu, L., Barber, A., Nuriel, S., Wagner, H.D.: Mechanical properties of functionalized SWCNT/PVA nanocomposites. Adv. Funct. Mater. **15**(6), 975–980 (2005)

Liu, L., et al.: Comparison of covalently and noncovalently functionalized carbon nanotubes in epoxy. Macromol. Rapid Commun. **30**(8), 627–632 (2009)

Liu, L., et al.: The effects of the variations of carbon nanotubes on the micro-tribological behavior of carbon nanotubes/bismaleimide nanocomposite. Compos. A **38**(9), 1957–1964 (2007a)

Liu, L.Q., et al.: One step electrospun nanofiber-based composite ropes. Appl. Phys. Lett. **9**(8), 083108 (2007b)

Liu, L.Q., Tasis, D., Prato, M., Wagner, H.D.: Tensile mechanics of electrospun multiwalled nanotube/poly(methyl methacrylate) nanofibers. Adv. Mater. **19**(9), 1228–1233 (2007c)

López Manchado, M.A., Valentini, L., Biagiotti, J., Kenny, J.M.: Thermal and mechanical properties of single-walled carbon nanotubes-polypropylene composites prepared by melt processing. Carbon **43**(7), 1499–1505 (2005)

Maccagno, T.M., Knott, J.F.: The fracture behaviour of PMMA in mixed modes I and II. Eng. Fract. Mech. **34**(1), 65–86 (1989)

Manoharan, M.P., et al.: The interfacial strength of carbon nanofiber epoxy composite using single fiber pullout experiments. Nanotechnology **20**(29), 295701 (2009)

Ma, P.C., Kim, J.K., Tang, B.Z.: Effects of silane functionalization on the properties of carbon nanotube/epoxy nanocomposites. Compos. Sci. Technol. **67**(14), 2965–2972 (2007)

Ma, P.C., Siddiqui, N.A., Marom, G., Kim, J.K.: Dispersion and functionalization of carbon nanotubes for polymer-based nanocomposites: a review. Compos. A **41**(10), 1345–1367 (2010)

Masuda, J., Torkelson, J.M.: Dispersion and major property enhancements in polymer/multiwall carbon nanotube nanocomposites via solid-state shear pulverization followed by melt mixing. Macromolecules **41**(16), 5974–5977 (2008)

McIntosh, D., Khabashesku, V.N., Barrera, E.V.: Nanocomposite fiber systems processed from fluorinated SWCNTs and a polypropylene matrix. Chem. Mater. **18**(9), 4561–4569 (2006)

McIntosh, D., Khabashesku, V.N., Barrera, E.V.: Benzoyl peroxide initiated in situ functionalization, processing and mechanical properties of SWCNT–polypropylene composite fibers. J. Phys. Chem. C **111**(4), 1592–1600 (2007)

Meincke, O., et al.: Mechanical properties and electrical conductivity of carbon-nanotube filled polyamide-6 and its blends with acrylonitrile/butadiene/styrene. Polymer **45**(3), 739–748 (2004)

Mikelson, E.T., et al.: Fluorination of single-wall carbon nanotubes. Chem. Phys. Lett. **296**(1–2), 188–194 (1998)

Mirjalili, V., Hubert, P.: Modelling of the carbon nanotube bridging effect on the toughening of polymers and experimental verification. Compos. Sci. Technol. **70**(10), 1537–1543 (2010)

Mittal, G., et al.: A review on carbon nanotubes and graphene as fillers in reinforced polymer nanocomposites. J. Ind. Eng. Chem. **21**, 11–25 (2015)

Miyagawa, H., Drzal, L.T.: Thermophysical and impact properties of epoxy nanocomposites reinforced by SWCNTs. Polymer **45**(15), 5163–5170 (2004)

Moore, E.M., et al.: Enhancing the strength of polypropylene fibers with CNTs. J. Appl. Polym. Sci. **93**(6), 2926–2933 (2004)

Ogasawara, T., Ishida, Y., Ishikawa, T., Yokota, R.: Characterization of multi-walled carbon nanotube/phenylethynyl terminated polyimide composites. Compos. A **35**(1), 67–74 (2004)

Paiva, M.C., et al.: Mechanical and morphological characterization of polymer–carbon nanocomposites from functionalized CNTs. Carbon **42**(14), 2849–2854 (2004)

Park, C., et al.: Dispersion of single wall carbon nanotubes by in situ polymerization under sonication. Chem. Phys. Lett. **364**(3–4), 303–308 (2002)

Park, J.M., Kim, D.S., Lee, J.R., Kim, T.W.: Nondestructive damage sensitivity and reinforcing effect of CNT/epoxy composites using electromicromechanical technique. Mater. Sci. Eng. C **23**(6–8), 971–975 (2003)

Pekker, S., et al.: Hydrogenation of carbon nanotubes and graphite in liquid ammonia. Phys. Chem. **105**(33), 7938–7943 (2001)

Potschke, P., Bhattacharyya, A.R., Janke, A., Goering, H.: Melt mixing of polycarbonate/multiwall carbon nanotube composites. Compos. Interf. **10**(4), 389–404 (2003)

Pötschke, P., Bhattacharyya, A.R., Janke, A.: Carbon nanotube-filled polycarbonate composites produced by melt mixing and their use in blends with polyethylene. Carbon **42**(5), 965–969 (2004)

Probst, O., Moore, E.M., Resasco, D.E., Grady, B.P.: Nucleation of PVA crystallization by SWCNTs. Polymer **45**(13), 4437–4443 (2004)

Putz, K.W., Mitchell, C.A., Krishnamoorti, R., Green, P.F.: Elastic modulus of single-walled carbon nanotube/poly(methyl methacrylate) nanocomposites. J. Polym. Sci. Part B: Polym. Phys. **42**(12), 2286–2293 (2004)

Qian, D., Dickey, E.C., Andrews, R., Rantell, T.: Load transfer and deformation mechanisms in carbon nanotube–polystyrene composites. Appl. Phys. Lett. **76**(20), 2868–2870 (2000)

Rahmanian, S., et al.: Growth of carbon nanotubes on silica microparticles and their effects on mechanical properties of polypropylene nanocomposites. Mater. Des. **69**, 181–189 (2014)

Ren, Y., Li, F., Cheng, H.M., Liao, K.: Tension–tension fatigue behavior of unidirectional single-walled carbon nanotube. Carbon **41**(11), 2159–2179 (2003)

Ruan, S.L., Gao, P., Yang, X.G., Yu, T.X.: Toughening high performance ultrahigh molecular weight polyethylene using multiwalled carbon nanotubes. Polymer **44**(19), 5643–5654 (2003)

Ruan, S., Gao, P., Yu, T.X.: Ultra-strong gel-spun UHMWPE fibers reinforced using multiwalled carbon nanotubes. Polymer **47**(5), 1604–1611 (2006)

Ryan, K.P., et al.: Carbon nanotubes for reinforcements of plastics? A case study with PVA. Compos. Sci. Technol. **67**(7–8), 1640–1649 (2007)

Sabba, Y., Thomas, E.L.: High-concentration dispersion of single-wall carbon nanotubes. Macromolecules **37**(13), 4815–4820 (2004)

Safadi, B., Andrews, R., Grulke, E.A.: Multiwalled carbon nanotube polymer composites: synthesis and characterization of thin films. J. Appl. Polym. Sci. **84**(14), 2660–2669 (2002)

Safdari, M., Al-Haik, M.S.: Synergistic electrical and thermal transport properties of hybrid polymeric nanocomposites based on carbon nanotubes and graphite nanoplatelets. Carbon **64**, 111–121 (2013)

Sandler, J.K.W., et al.: Ultra-low electrical percolation threshold in carbon-nanotube-epoxy composites. Polymer **44**(19), 5893–5899 (2003)

Santangelo, S., et al.: Polylactide and carbon nanotubes/smectite-clay nanocomposites: preparation, characterization, sorptive and electrical properties. Appl. Clay Sci. **53**(2), 188–194 (2011)

Satapathy, B.K., Weidisch, R., Pötschke, P., Janke, A.: Tough-to-brittle transition in multiwalled carbon nanotube (MWNT)/polycarbonate nanocomposites. Compos. Sci. Technol. **67**(5), 867–879 (2007)

Schmid, C.F., Klingenberg, D.J.: Mechanical flocculation in flowing fiber suspensions. Phys. Rev. Lett. **84**(2), 290–293 (2000)

Seyhan, A.T., Tanoğlu, M., Schulte, K.: Tensile mechanical behavior and fracture toughness of Mwcnt and Dwcnt modified vinyl-ester/polyester hybrid nanocomposites produced by 3-roll milling. Mater. Sci. Eng. A **523**(1–2), 85–92 (2009)

S-Hadavand, B., Mahdavı Javıd, K., Gharagozlou, M.: Mechanical properties of multi-walled carbon nanotube/epoxy polysulfide nanocomposite. Mater. Des. **50**, 62–67 (2013)

Shadlou, S., Alishahi, E., Ayatollahi, M.R.: Fracture behavior of epoxy nanocomposites reinforced with different carbon nano-reinforcements. Compos. Struct. **95**, 577–581 (2013)

Shaffer, M.S.P., Windle, A.H.: Fabrication and characterization of carbon nanotube/poly(vinyl alcohol) composites. Adv. Mater. **11**(11), 937–941 (1999)

Shen, X.J., Pei, X.Q., Liu, Y., Fu, S.Y.: Tribological performance of carbon nanotube–graphene oxide hybrid/epoxy composites. Compos. Part B **57**, 120–125 (2014)

Shim, M., Wang, C., Guyot-Sionnest, P.: Charge-tunable optical properties in colloidal semiconductor nanocrystals. Phys. Chem. B **105**(12), 2369–2373 (2001)

Shofner, M.L., Khabashesku, V.N., Barrera, E.V.: Processing and mechanical properties of fluorinated SWCNT–polyethylene composites. Chem. Mater. **18**(4), 906–913 (2006)

Shokrieh, M.M., Kefayati, A.R., Chitsazzadeh, M.: Investigation of mechanical properties of clay/epoxy nanocomposites. In: Kish Island-Iran, the 2nd International Conference on Composites: Characterization, Fabrication & Application (CCFA-2) (2010)

Shokrieh, M.M., Esmkhani, M., Haghighatkhah, A.R., Zhao, Z.: Flexural fatigue behavior of synthesized graphene/carbon-nanofiber/epoxy hybrid nanocomposites. Mater. Des. **62**, 401–408 (2014)

Spitalsky, Z., Krontiras, C.A., Georga, S.N., Galiotis, C.: Effect of oxidation treatment of multi-walled carbon nanotubes on the mechanical and electrical properties of their epoxy composites. Compos. A **40**(6–7), 778–783 (2009a)

Spitalsky, Z., et al.: Modification of carbon nanotubes and its effect on properties of carbon nanotube/epoxy nanocomposites. Polym. Compos. **30**(10), 1378–1387 (2009b)

Spitalsky, Z., Tasis, D., Papagelis, K., Galiotis, C.: Carbon nanotube–polymer composites: chemistry, processing, mechanical and electrical properties. Prog. Polym. Sci. **56**(3), 357–401 (2010)

Sui, G., Zhong, W.H., Liu, M.C., Wu, P.H.: Enhancing mechanical properties of an epoxy resin using "liquid nano-reinforcements". Mater. Sci. Eng. A **512**(1–2), 139–142 (2009)

Sulong, A.B., et al.: Electrical conductivity behaviour of chemical functionalized MWCNTs epoxy nanocomposites. Eur. J. Sci. Res. **59**(1), 13–21 (2009)

Sun, Y.P., et al.: Soluble dendron-functionalized carbon nanotubes: preparation, characterization. Chem. Mater. **13**(9), 2864–2869 (2001)

Sun, Y.P., Fu, K., Lin, Y., Huang, W.: Functionalized carbon nanotubes: properties and applications. Acc. Chem. Res. **35**(12), 1096–1104 (2002)

Sun, L., et al.: Mechanical properties of surface-functionalized SWCNT/epoxy composites. Carbon **46**(2), 320–328 (2008)

Switzer, L.H., Klingenberg, D.J.: Flocculation in simulations of sheared fiber suspensions. Int. J. Multiph. Flow **30**(1), 67–87 (2004)

Tang, W., Santare, M.H., Advani, S.G.: Melt processing and mechanical property characterization of multi-walled carbon nanotube high density polyethylene composite films. Carbon **41**(14), 2779–2785 (2003)

Thostenson, E.T., Chou, T.W.: Aligned multi-walled carbon nanotube reinforced composites: processing and mechanical characterization. J. Phys. D Appl. Phys. **35**(16), 77–80 (2002)

Thostenson, E.T., Chou, T.W.: Processing–structure–multi-functional property relationship in carbon nanotube/epoxy composites. Carbon **44**(14), 3022–3029 (2006)

Tong, X., et al.: Surface modification of single-walled carbon nanotubes with polyethylene via in situ Ziegler-Natta polymerization. J. Appl. Polym. Sci. **92**(6), 3697–3700 (2004)

Tong, L., Sun, X., Tan, P.: Effect of long multi-walled carbon nanotubes on delamination toughness of laminated composites. J. Compos. Mater. **42**(1), 5–23 (2008)

Tseng, C.H., Wang, C.C., Chen, C.Y.: Functionalizing carbon nanotubes by plasma modification for the preparation of covalent-integrated epoxy composites. Chem. Mater. **19**(2), 308–315 (2007)

Valentini, L., et al.: Use of plasma fluorinated single-walled carbon nanotubes for the preparation of nanocomposites with epoxy matrix. Compos. Sci. Technol. **68**(3–4), 1008–1014 (2008)

Velasco-Santos, C., et al.: Dynamical–mechanical and thermal analysis of carbon nanotube–methyl-ethyl methacrylate nanocomposites. J. Phys. D Appl. Phys. **36**(12), 1423–1428 (2003a)

Velasco-Santos, C., et al.: Improvement of thermal and mechanical properties of carbon nanotube composites through chemical functionalization. Chem. Mater. **15**(23), 4470–4475 (2003b)

Vigolo, B., et al.: An experimental approach to the percolation of sticky nanotubes. Science **309**(5736), 920–923 (2005)

Wang, Q., Pei, X.: The influence of nanoparticle fillers on the friction and wear behavior of polymer matrices. Tribology of Polymeric Nanocomposites, pp. 63–64. Elsevier Science, s.l. (2008)

Wang, Z., et al.: Processing and property investigation of single-walled carbon nanotube (SWNT) buckypaper/epoxy resin matrix nanocomposites. Compos. A **35**(10), 1225–1232 (2004)

Wang, Y., Cheng, R., Liang, L., Wang, Y.: Study on the preparation and characterization of ultra high molecular weight polyethylene–carbon nanotubes composite fiber. Compos. Sci. Technol. **65**(5), 793–797 (2005)

Wang, M., Pramoda, K.P., Goh, S.H.: Enhancement of interfacial adhesion and dynamic mechanical properties of PMMA–MWCNT composites with amine-terminated poly(ethylenoxide). Carbon **44**(4), 613–617 (2006a)

Wang, S., et al.: Effective functionalization of carbon nanotubes for reinforcing epoxy polymer composites. Nanotechnology **17**(6), 1551–1557 (2006b)

Wang, C., et al.: A study on microhardness and tribological behaviour of carbon nanotubes reinforced AMMA-CNTs copolymer nanocomposites. Mater. Sci. Eng. A **478**(1–2), 314–318 (2008)

Weisenberger, M.C., et al.: Enhanced mechanical properties of polyacrylonitrile/multiwall carbon nanotube composite fibers. J. Nanosci. Nanotechnol. **3**(6), 535–539 (2003)

Wetzel, B., Rosso, P., Haupert, F., Friedrich, K.: Epoxy nanocomposites—fracture and toughening mechanisms. Eng. Fract. Mech. **73**(16), 2375–2398 (2006)

Witt, N., Tang, Y., Ye, L., Fang, L.: Silicone rubber nanocomposites containing a small amount of hybrid fillers with enhanced electrical sensitivity. Mater. Des. **45**, 548–554 (2013)

Xia, H., Wang, Q., Li, K., Hu, G.H.: Preparation of CNT/polypropylene composite powder with a solid state mechanochemical pulverization process. J. Appl. Polym. Sci. **93**(1), 378–386 (2004)

Xiao, K.Q., Zhang, L.C., Zarudi, I.: Mechanical and rheological properties of CNT–reinforced polyethylene composites. Compos. Sci. Technol. **67**(2), 177–182 (2007a)

Xiao, Y., et al.: Dispersion and mechanical properties of polypropylene/multiwall carbon nanotubes composites obtained via dynamic packing injection molding. J. Appl. Polym. Sci. **104**(3), 1880–1886 (2007b)

Xie, L., et al.: Single-walled carbon nanotubes functionalized with high bonding density of polymer layers and enhanced mechanical properties of composites. Macromolecules **40**(9), 3296–3305 (2007)

Xu, X., Thwe, M.M., Shearwood, C., Liao, K.: Mechanical properties and interfacial characteristics of carbon-nanotube–reinforced epoxy thin films. Appl. Phys. Lett. **81**(15), 2833–2835 (2002)

Yang, B.X., Pramoda, K.P., Xu, G.Q., Goh, S.H.: Mechanical reinforcement of polyethylene using polyethylene-grafted multiwalled carbon nanotubes. Adv. Funct. Mater. **17**(13), 2062–2069 (2007)

Yang, K., et al.: Effects of carbon nanotube functionalisation on the mechanical and thermal properties of epoxy composites. Carbon **47**(7), 1723–1737 (2009)

Zhang, X., et al.: Poly(vinyl alcohol)/SWCNT composite film. Nano Lett. **3**(9), 1285–1288 (2003)

Zhang, X., et al.: Gel spinning of PVA/SWCNT composite fiber. Polymer **45**(26), 8801–8807 (2004)

Zhang, W., Picu, R.C., Koratkar, N.: The effect of carbon nanotube dimensions and dispersion on the fatigue behavior of epoxy nanocomposites. Nanotechnology **19**(28), 285709 (2008)

Zhao, P., et al.: Excellent tensile ductility in highly oriented injection-molded bars of polypropylene/carbon nanotubes composites. Polymer **48**(19), 5688–5695 (2007)

Zhou, Y., Pervin, F., Lewis, L., Jeelani, S.: Fabrication and characterization of carbon/epoxy composites mixed with multi-walled carbon nanotubes. Mater. Sci. Eng. A **475**(1–2), 230–237 (2008)

Zou, Y., Feng, Y., Wang, L., Liu, X.: Processing and properties of MWNT/HDPE composites. Carbon **42**(2), 271–277 (2004)

Chapter 5
CNT/FRP Composites

5.1 Introduction

The use of fiber reinforced polymer composites (FRPs) is enhancing in different engineering applications because of their high specific strength and stiffness. However, the through-the-thickness behavior of FRPs lacks some of the important mechanical property requirements for structural applications, such as airplane components or marine structures. Some researchers have recommended the addition of nanoparticles like carbon nanotubes and carbon nanofibers (CNFs) for improving the FRP properties in the thickness direction with very little increase in the weight of composite.

The carbon-based nanoreinforcements are often added to the material to improve its multi-functionality and to customize its mechanical, electrical and thermal properties. While most of the recent studies have dealt with the influence of nanoreinforcements on the polymer matrix, few researchers have investigated composites that are strengthened in multi-scale. As an example for such multiscale composites, one may mention the nano-enhancement of classical laminated composites, in particular carbon-fiber/epoxy-resin laminated composites.

As described by Lubineau and Rahaman (2012), carbon fiber-based laminated composites are known as popular materials in secondary structures in aeronautical applications (i.e. moderately loaded non-critical structures). More recently, these composite materials have been used in critical components, such as the central beam or the central wing box of an airplane with the aim to increase the proportion of composite materials in the next-generation aircrafts. While these composites have reached their limit in terms of mechanical properties, some initial studies have been conducted on how to strengthen them by nanoparticle reinforcement.

Investigations dealing with nano-structured materials often report new material formations and assess them according to few separate properties rather than on

---

The original version of this chapter was revised. The Erratum to this chapter is available at 10.1007/978-3-319-00251-4_7

M. Rahmandoust and M.R. Ayatollahi, *Characterization of Carbon Nanotube-Based Composites Under Consideration of Defects*, Advanced Structured Materials 39,
DOI 10.1007/978-3-319-00251-4_5

the complete set of working mechanisms in real composite structures. Structural modeling techniques have helped researchers to identify well the concepts of degradation in classical laminated composites. The valuable knowledge on the mechanisms of degradation in classical composites provides a background for identifying and understanding the likely defects in nanoreinforced FRPs.

The first part of this chapter focuses on the processing techniques of multiscale composites containing carbon nanotubes or carbon nanofibers. In the second part, the mechanical properties of polymer-based composites reinforced by CNT are studied. Finally, in the last part, a limited number of studies that deal with the tribological properties of these multiscale composites are described.

5.2 Production Methods

The exploitation of various properties of CNTs in a composite depends strongly on how the CNTs are integrated into the three-phase multiscale composites. In general, when discussing polymer/CNT composites, two major classes come to mind. First, the CNT nano-fillers are dispersed within a polymer at a specified concentration, and the entire mixture is fabricated into a composite. Secondly, as-grown CNT are processed into fibers or films, and this macroscopic CNT material is then embedded into a polymer matrix (Song et al. 2013). Modifying the matrix resin by dispersing CNTs in the polymer offers the most conventional, facile and industrially compatible route. A typical scanning electron microscope (SEM) image of the three-phase CNT-carbon fiber FRP (CFRP) multiscale composite fabricated by modifying the matrix resin with CNTs is presented in Fig. 5.1 (Yokozeki et al. 2007).

A number of different techniques have been developed for CNT dispersion into a polymer matrix, which includes shear mixing, calendaring, extrusion, ultrasonication and ball milling (Khan and Kim 2011). Many of the recent studies are often

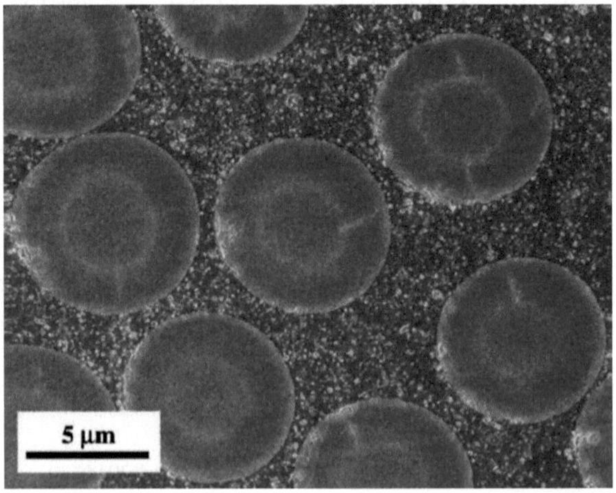

Fig. 5.1 A sample SEM picture of a multiscale carbon nanotube-carbon fiber reinforced polymer composite (Yokozeki et al. 2007)

based on the use of a combination of aforementioned techniques, such as ultrasonication plus ball milling (Ma et al. 2009), and ultrasonication plus extrusion (Isayev et al. 2009).

Fabrication of multiscale FRPs using modified resins has some limitations that hinder their development and widespread applications. One of the main problems in this approach is that the viscosity of a CNT-modified matrix increases dramatically with increasing CNT content, even with a very small content below 1 wt%, leading to severe agglomeration of CNTs in the bulk of the composite. The highly viscous and CNT-agglomerated resin systems cannot be processed using the conventional composite manufacturing techniques (Khan and Kim 2011). For example, when fabricating multiscale FRPs using the vacuum assisted resin transfer molding (VARTM) process, nanofillers can become filtered, resulting in an inhomogeneous microstructure of multiscale composites (Fan et al. 2008; Green et al. 2009; Sadeghian et al. 2006; Qiu et al. 2007). Filtering is caused by the nanofillers being trapped in inter-tow regions within the fiber preform mesh during fabrication (Khan and Kim 2011). A comprehensive study revealed a direct correlation between the concentration of CNFs and the processing parameters, such as viscosity and filtering: a CNF concentration above 1 % resulted in a significant filtering effect, leading to the formation of many microvoids in the composite (Sadeghian et al. 2006).

Modifying the fibers includes chemical vapor decomposition or electrophoresis techniques. Various techniques, such as electric arc discharge, laser ablation, chemical vapor deposition, etc., are used to produce CNTs. Chemical vapor decomposition (CVD) is the only method that can be used to grow carbon nanofillers (CNTs, CNFs, CNCs) on a fiber surface (Fig. 5.2).

CVD synthesis of carbon nanofillers is achieved by thermal dissociation of carbon-containing gaseous molecules at high temperature followed by the deposition of carbon on a catalyst-coated fiber substrate. Catalysts are used to facilitate both the dissociation of carbon source molecules in an active radical species and

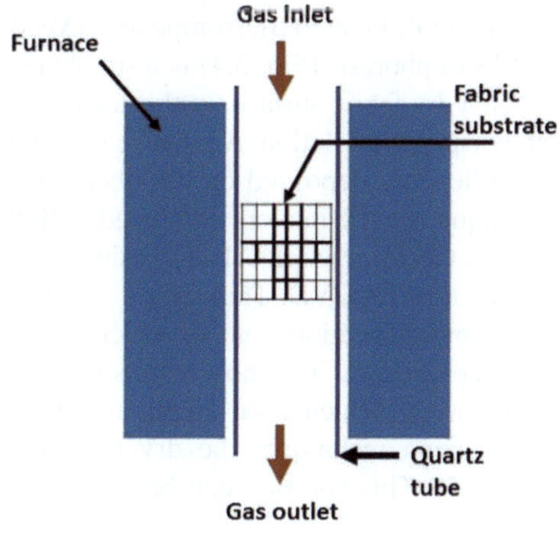

Fig. 5.2 Typical procedure for chemical vapor decomposition (CVD) process

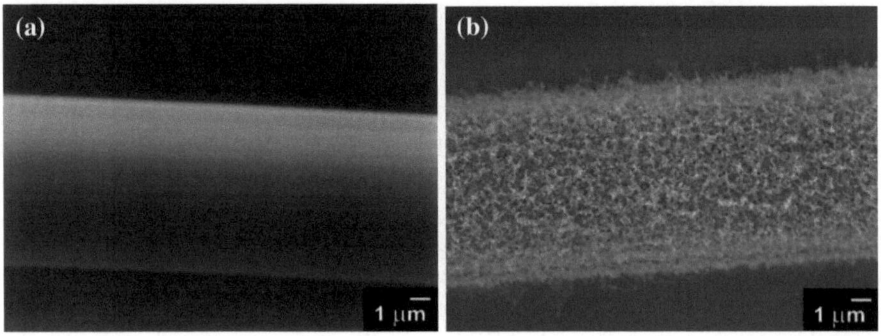

Fig. 5.3 Sample SEM micrographs of carbon fibers **a** before, and **b** after carbon nanotube growth (Thostenson et al. 2002) (Reprinted with permission from Thostenson et al. Copyright 2002, AIP Publishing LLC)

the growth of carbon nanofillers. Commonly used gaseous carbon sources include methane, acetylene and carbon monoxide, etc. (Lubineau and Rahaman 2012). Thostenson et al. (2002) were the first group of researchers who successfully synthesized CNTs via chemical vapor deposition (CVD) on the surface of carbon fibers as shown in Fig. 5.3. The effects of growth conditions on CNT morphology have been evaluated in the subsequent studies (Khan and Kim 2011).

Shen et al. grew CNTs directly on a (Fe)-coated carbon fiber catalyst by CVD in a hydrocarbon atmosphere in the temperature range of 550–1000 °C (Zhu et al. 2003). Otsuka et al. reported winding-type CNTs on Ni/VGCF or Ni/GCF by catalytic decomposition of benzene and cyclopentadiene with uniform diameters (10–20 nm) at a relatively low temperature (500 °C), while the fishbonetype CNTs are formed from catalytic decomposition of methane and n-hexane (Otsuka et al. 2004). Ismagilov et al. have observed random and non-uniform growth of carbon nanofibers on silica glass fibers (Ismagilov et al. 2005). Veedu et al. grew CNTs perpendicular to 2D woven fabrics of SiC by CVD method. These fibers are then impregnated by a high-temperature epoxy matrix and are subsequently stacked to prepare multilayered 3D composites (Veedu et al. 2006).

Electrophoresis (Fig. 5.4) is a simple technique for deposition of carbon nanofillers on a dry fabric surface. In this technique, the carbon fabric is dipped in the nanotube dispersion solution. A positive potential is applied to the fabric such that carbon nanofillers are deposited on the fiber surface. The CNT coating obtained with this technique is quite random compared with that obtained by CVD, which can achieve aligned CNTs networks. The density of the CNTs on the fiber surface is generally smaller than for CVD. This carbon nanofiller coated fiber is expected to have various important applications, one of which is as a reinforcing material for preparing laminated composites. The most interesting point in this technique is clearly its scalability and its implementation simplicity (Lubineau and Rahaman 2012).

In sizing technique, the dry fiber is coated with a CNTs-enriched solution (Fig. 5.5). This solution can be based on a classical commercial sizing agent, or

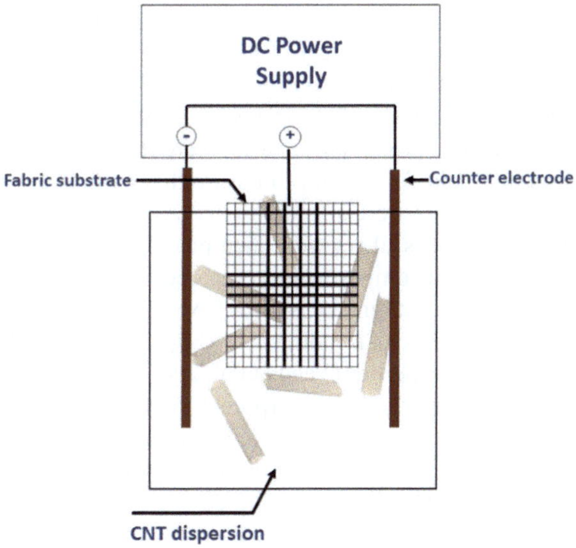

Fig. 5.4 Schematics of the electrophoresis process

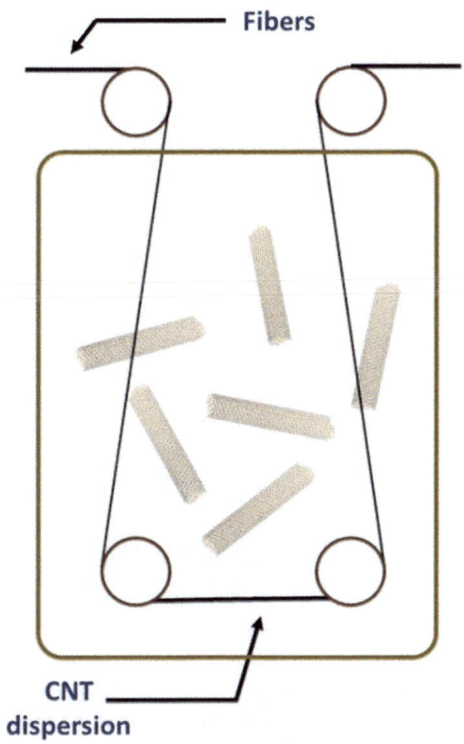

Fig. 5.5 Schematics of the sizing process

on a polymer matrix that will be deposited as a thin layer over the fiber surface (Lubineau and Rahaman 2012). Barbar et al. presented a method for integrating CNTs on glass fibers via mixing of CNTs with an industrial sizing agent. First, they dispersed SWCNTs in a sizing solution at a concentration of 0.5 wt% using an ultrasonic probe. Glass fibers were then immersed in the sizing-SWCNTs solution and held for 10 min to ensure that each fiber was fully covered with the sizing. The fibers were allowed to dry vertically at room temperature for 1 week. The SWCNTs-glass fibers were then placed between two sheets of polypropelene and a hot-pressed composite was prepared (Barber et al. 2004).

Grafting of aligned CNTs or CNFs on to the conventional fiber surface resulted in a 3D hierarchical composite (Fig. 5.6) with enhanced mechanical and multifunctional properties (Veedu et al. 2006).

Another method to incorporate the CNTs into the composite is reinforcements for interlaminar modification in prepreg-based composites. Interestingly, much less has been published on the use of nanofillers as reinforcements for interlaminar modification in prepreg-based composites although this strategy has attracted increasing interest in the recent years. The processing techniques can be classified depending on the final arrangements of the nanoreinforcements over the interface, which can be either random or aligned (Lubineau and Rahaman 2012).

To achieve random dispersion, the main techniques are powder dispersion, spraying, roll coating and electrospinning (Lubineau and Rahaman 2012). A technique known as 'powder method' was devised to add CNTs onto the ply surface

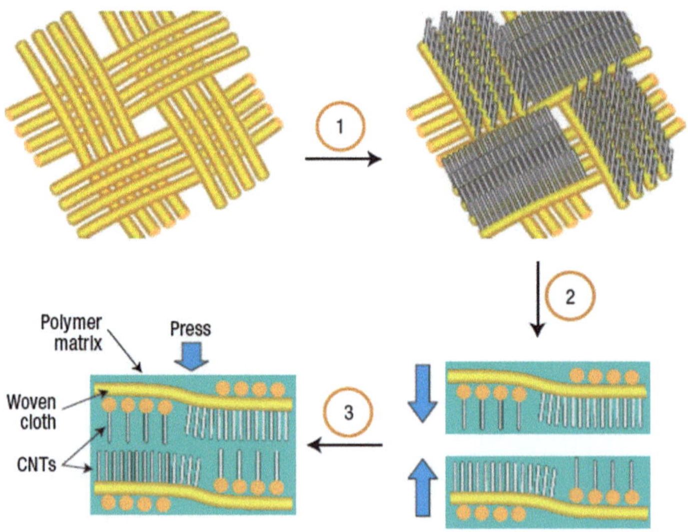

Fig. 5.6 Typical steps involved in the hierarchical manufacturing of a three-dimensional composite: (1) aligned nanotube grown on the fiber cloth; (2) stacking of matrix-infiltrated carbon nanotube-grown fiber cloth; (3) 3D nanocomposite plate fabrication by hand lay-up (Veedu et al. 2006) (Reprinted by permission from Macmillan Publishers Ltd: *Nature Materials* Veedu et al. copyright 2006)

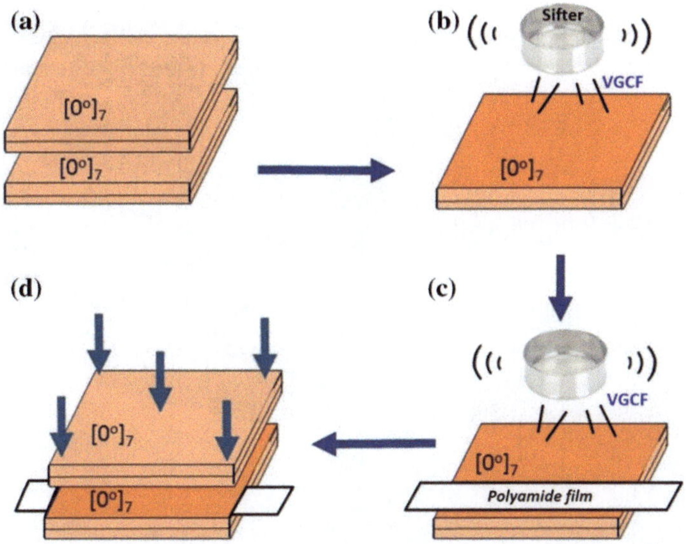

Fig. 5.7 Schematics of powder method for dispersing vapor grown carbon fiber between carbon fiber reinforced polymer composite prepregs

(Li et al. 2009). The method is schematically described in Fig. 5.7 where the powder of vapor grown carbon fibers (VGCFs) are manually spread at the mid-plane of CFRP laminates using a sifter with a mesh size about 70 μm. A VGCF interlayer is automatically formed during melting and curing of resin at a high temperature and pressure in the autoclave. A CNT-solvent paste, e.g. CNF ethanol paste, was also applied manually on a CFRP prepreg using a metal roller (Arai et al. 2008). After evaporating the solvent on the prepreg sheets, the two parts of prepregs were bonded together and cured. A CNF interlayer of 50–200 μm in thickness was naturally formed by the fusion of CNFs and epoxy resin during curing. A thin layer of CNTs was also dispersed onto the surface of woven fabric using a spray-up process (Zhu et al. 2007). This process involved coating of a pre-dispersed CNT-solvent solution on the fabric using a spray gun, and deposition of CNTs at the prescribed weight percentage on the fabric after evaporation of the solvent. Composites were then produced through a RTM process. CNTs were also integrated into FRPs by adding CNTs in the fiber sizing formulation through which the continuous fibers were drawn (Warrier et al. 2010). Mujika et al. dispersed the CNTs over the interface by spraying a solvent dispersion of CNTs with an airbrush. When using this technique, the solvent needs to be evaporated before poling the prepregs together (Mujika et al. 2012).

To control the alignment of the nanoreinforcements, Garcia et al. reinforced the interlaminar interface between the prepregs by depositing a forest of aligned CNTs. This was accomplished by growing a vertically-aligned CNT forest at a high temperature, and then transfer-printing the CNTs to prepregs at room temperature. The prepreg was attached to a cylindrical drum that was rolling under

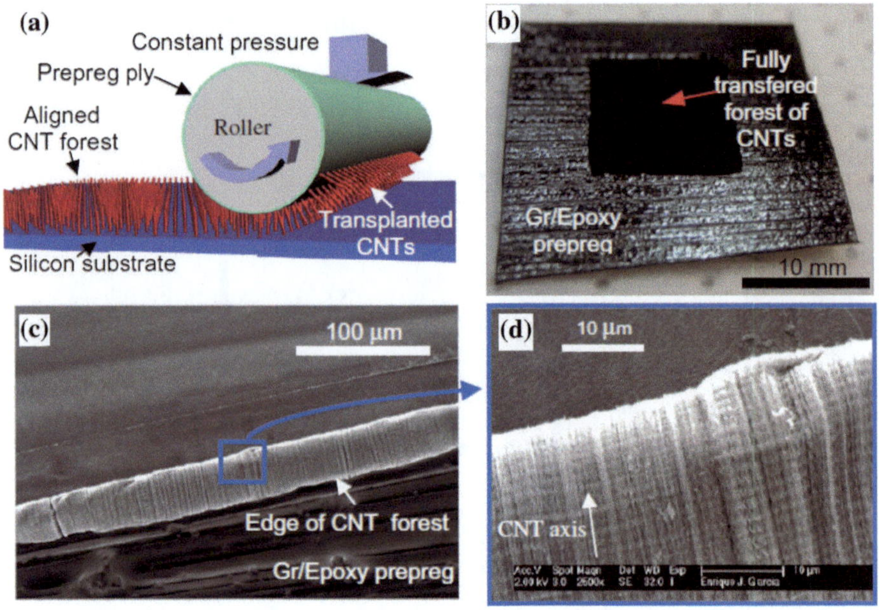

Fig. 5.8 The procedure for transfer-printing of vertically-aligned carbon nanotubes (VACNTs) to prepreg: **a** illustration of the 'transfer-printing' process; **b** CNT forest fully transplanted from its original silicon substrate to the surface of prepreg ply; (**c** and **d**) SEM micrographs of CNT forest, showing CNT alignment after transplantation (Garcia et al. 2008b)

pressure across the substrate containing the CNT forest to transfer the CNTs to prepregs, taking advantage of the tack of the prepregs to separate the CNT forest from the growth substrate (Fig. 5.8) (Garcia et al. 2008b).

In summary, fibers and the bulk matrix can be used with any processing technique with only raw materials, such as the hand lay-up technique, resin transfer molding (RTM) and the vacuum impregnation technique. Reinforcing the interlaminar interface is the only possible choice when the processing technique is based on intermediate products such as prepregs (Lubineau and Rahaman 2012). Figure 5.9 shows a flow chart for the processing techniques of nanoreinforced laminated composites.

5.3 Mechanical Properties

5.3.1 Effects of Matrix Modification

Siddiqui et al. evaluated the effects of carbon nanotube (CNT)–epoxy nanocomposite coating applied to glass fiber surface on tensile strength of single glass fibers at different gauge lengths. The results indicated that the tensile strength of

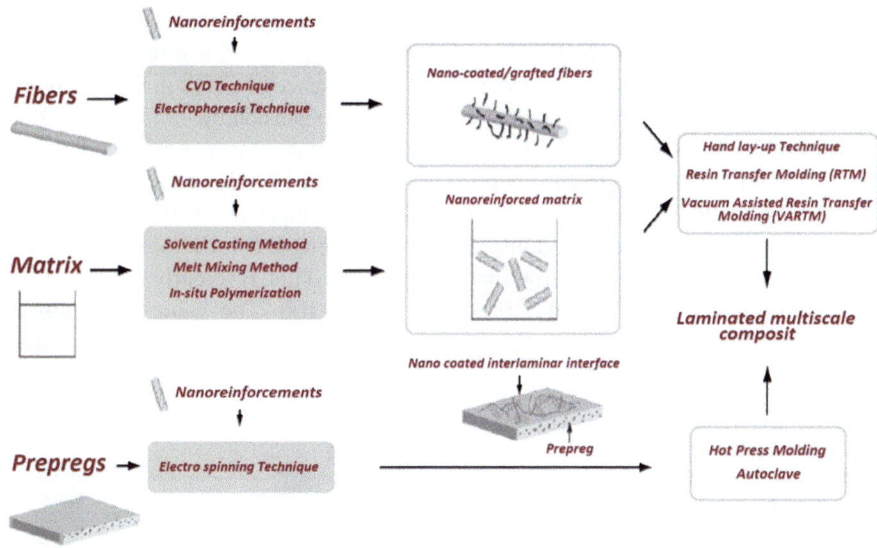

Fig. 5.9 Typical stages for the processing techniques of nanoreinforced laminated composites

single fibers increased significantly with increasing CNT content up to a certain level, depending on the type of CNTs. The crack healing efficiency was much higher for the fibers coated with straight, less entangled CNTs than those with highly entangled CNTs, indicating that the CNT dispersion state in the coating played an important role. The coated individual glass fibers were mounted on specially prepared cardboard frames according to the specifications (ASTM Standard D3822-1). Specimens with different gauge lengths of 25, 50, 100 and 150 mm gauge lengths were prepared. Table 5.1 shows tensile strengths of fibers with and without coatings measured at different gauge lengths (Siddiqui et al. 2010).

Godara et al. studied the influence of carbon nanotube reinforcement on the mechanical properties of carbon fiber/epoxy composites. They dispersed a fixed amount (0.5 wt%) of different types of CNTs (functionalized and non-functionalized) in the epoxy

Table 5.1 Tensile strengths of fibers with and without coatings measured at different gauge lengths (mean value ± one std) (Siddiqui et al. 2010)

Type of coating	25 mm (MPa)	50 mm (MPa)	100 mm (MPa)	150 mm (MPa)
No coating	2405 ± 275	1890 ± 264	1604 ± 165	1205 ± 205
Epoxy coating	2498 ± 280	2077 ± 260	1853 ± 190	1536 ± 160
0.3 % VNT–epoxy coating	2577 ± 215	2172 ± 207	2057 ± 295	1643 ± 155
0.5 % VNT–epoxy coating	2000 ± 285	–	1364 ± 200	–
0.3 % BNT–epoxy coating	2599 ± 270	2232 ± 225	1831 ± 221	1684 ± 226
0.5 % BNT–epoxy coating	2785 ± 240	2597 ± 207	2305 ± 203	1802 ± 206
1.0 % BNT–epoxy coating	2392 ± 300	–	1612 ± 202	–

matrix, and produced unidirectional prepregs. The temperature–viscosity profile of the epoxy matrix reinforced with different types of CNTs indicated a strong dependency on the type of CNTs. The pronounced effect of the presence of CNTs in the matrix was reflected by the decrease of the coefficient of thermal expansion by ~32 % for the double-walled CNTs epoxy system. There was also a substantial increase in mode-I fracture toughness by over 80 % for the pristine multi-walled CNTs in combination with the epoxy resin modified by using a compatibilizer (Godara et al. 2009).

The longitudinal tensile modulus and tensile strength for UD composite laminates in general show fiber dominated behavior. This is clearly observed for the longitudinal tensile modulus plotted in Fig. 5.10b. The values are close to the value expected for 60 % carbon fiber (CF) volume fraction as predicted by

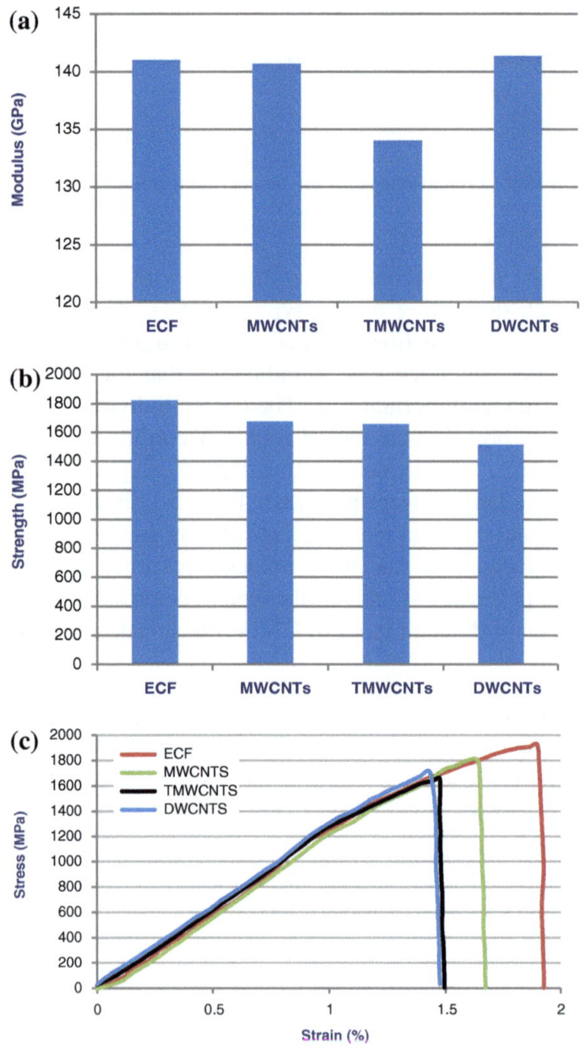

Fig. 5.10 Experimental results obtained for the tensile behavior of various UD-laminates tested in the fiber direction. **a** The tensile modulus, **b** failure strength, **c** stress versus strain curves (Godara et al. 2009)

the classical mixture rule. However, a systematic decrease is seen for the longitudinal tensile strength of up to 20 % (Fig. 5.10). This can be explained by the increased stress build-up around the broken fibers attributed to the stiffening effect of CNT on the effective matrix. Also CNT can act as defects and therefore degrade certain properties (in this case strength along the fiber direction), and at the same time be beneficial for other properties (like ductility under shear deformation) (Godara et al. 2009).

Chen et al. fabricated cross-ply laminates reinforced with basalt fibers and functionalized multi-walled carbon nanotubes (MWCNTs) from unidirectional epoxy prepregs. MWCNTs with varied surface conditions were prepared by oxidization or esterification, and then dispersed into a DGEBA epoxy system. The dispersion of the MWCNTs in the epoxy was improved by surface modification, resulting in improved composite mechanical properties as well. When MWCNT—filled epoxies were used as matrices for basalt fiber/epoxy laminates, however, the reinforcement effects of MWCNTs on the composite elastic modulus exceeded micromechanics based semi-empirical predictions and were independent of surface functionalization. Because of strong interfacial bonding between the PGE-CNT and the epoxy matrix, CNT debonding along the sidewall at the CNT matrix interface was not observed. The scarcity of the PGE-CNTs observed on basalt fiber surfaces was attributed to the alignment of PGE-CNTs along the fiber direction. Although the exact cause of the alignment is not well understood and awaits further investigation, CNT surface functionalities and shear flow within micron-sized inter-fiber channels during fabrication must play important roles in aligning and dispersing the PGE-CNTs. Additional evidence for the alignment of PGE-CNTs comes from the measured mechanical properties. Results of tensile tests for the multi-scale composites laminates are summarized in Table 5.2 (Chen et al. 2009).

Zhou et al. studied the effectiveness of adding CNTs on the mechanical and thermal properties of carbon/epoxy composites mixed with multi-walled carbon nanotubes. In this study, a high-intensity ultrasonic liquid processor was used to obtain a homogeneous mixture of epoxy resin and multi-walled carbon nanotubes (CNTs). The CNTs were infused into EPON 862 epoxy resin through sonic

Table 5.2 The values of Young's modulus and tensile strength obtained for multi-scale composite laminates (Chen et al. 2009)

Epoxy/basalt laminates	Young's modulus (GPa)	Improvement (%)	Tensile strength (MPa)	Improvement (%)
Control laminate 0 vol%	27.65 ± 0.41	0	584.7 ± 10.3	0
Raw CNT 0.5 vol%	27.44 ± 0.76	−0.76	564.0 ± 31.0	−3.54
o-CNT 0.5 vol%	30.41 ± 1.31	+9.98	635.7 ± 33.8	+8.72
PGE-CNT 0.5 vol%	29.92 ± 0.83	+8.20	608.8 ± 20.0	+3.98
Raw-CNT 1.5 vol%	28.53 ± 1.18	+3.73	504.0 ± 42.0	−13.8
o-CNT 1.5 vol%	36.4 ± 0.97	+31.65	627.7 ± 25.5	+7.35
PGE-CNT 1.5 vol%	34.90 ± 0.77	+26.2	615.1 ± 19.7	+5.20

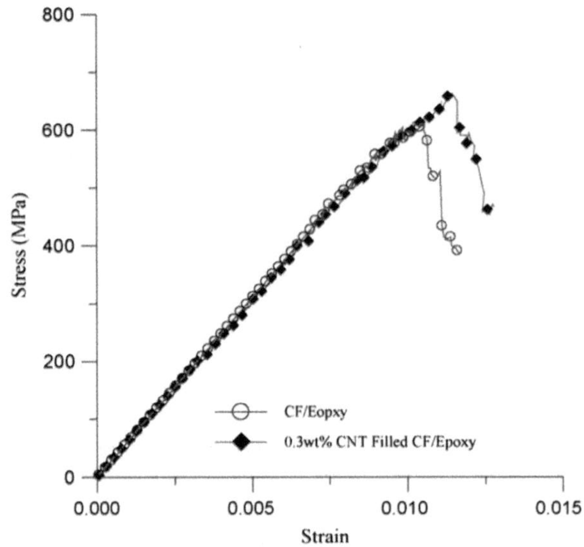

Fig. 5.11 Stress–strain curves obtained for neat and nanophased laminates in the experiments (Zhou et al. 2008)

cavitation and then mixed with W curing agent using a high-speed mechanical agitator. The trapped air and reaction volatiles were removed from the mixture using a high vacuum. They performed flexural tests and fracture toughness tests on unfilled and CNT-filled epoxy to identify the effect of adding CNTs on the mechanical properties of epoxy. The highest improvement in strength and fracture toughness was obtained with 0.3 wt% CNT loading. The nanophased matrix filled with 0.3 wt% CNT was then used with weave carbon fabric in a vacuum-assisted resin transfer molding (VARTM) set up to fabricate composite panels (Zhou et al. 2008).

Typical stress–strain curves of neat and nanophased laminates in this study are shown in Fig. 5.11. The curves show considerable non-linear deformation, and the irregularities in the curves were attributed to random filament breakage during loading. The specimens failed rapidly after reaching the maximum stress. The "pinging" noises were detected during the breakage of individual filaments, and during the inter-layer delamination. Stress–strain curves shown in Fig. 5.11 also reveal that, by infusing 0.3 wt% CNTs, strength and failure strain improve. The nanophased system showed an approximately 2 % increase in the flexural strength and a 3 % increase in the failure strain, compared to the neat system (Zhou et al. 2008).

Two possible reasons could be mentioned for the higher mechanical properties of the carbon fabric-reinforced laminate infused with CNTs. First, the increased strength of the epoxy matrix caused by CNTs, can increase the strength. Second, the presence of CNTs increases the crack propagation resistance by bridging effects, which improve the interlaminar strength (Zhou et al. 2008).

Fan et al. examined the interlaminar shear strength (ILSS) of traditional glass fiber reinforced epoxy composites enhanced by strategically injecting multi-walled carbon nanotube (MWCNT)–epoxy suspensions into stationary glass fiber mats. They prepared the suspensions by combining the techniques of high-speed mechanical stirring, ultrasonic agitation and acid oxidation. The results showed that the introduction of MWCNT into the composite could increase the ILSS by up to 33 % (Fan et al. 2008).

Results showed that the interlaminar shear strengths of OMWCNT/glass/epoxy composites in different weight percentages of 0, 0.5 and 1 wt% nanotubes were 32.49, 33.53 and 38.40 MPa, respectively. The average ILSS measured by short beam shear (SBS) technique improved 3.2 % by adding 0.5 wt% OMWCNT in epoxy, and 18.2 % for 1 wt% OMWCNT (Fan et al. 2008).

Fan et al. also fabricated four types of hybrid composite samples by infusing eight layers of 5 × 5 cm woven glass fiber mats with 0, 0.5, 1 and 2 wt% OMWCNT/epoxy suspensions by using the injection and double vacuum assisted resin transfer molding (IDVARTM) method. The results show that 0.5, 1, 2 wt% OMWCNT in epoxy increased the ILSS by 9.7, 20.5 and 33.1 %, respectively, with respect to the neat epoxy (0 wt%) (Fan et al. 2008).

Böger et al. evaluated the fatigue properties of glass fiber reinforced epoxy laminates modified with small amounts (0.3 wt%) multi-wall carbon nanotubes (MWCNT) by means of static (90°-tensile and stepped tensile) and dynamic fatigue tests. The addition of MWCNT lead to increases in the inter fiber fracture strength of up to 16 %. More significantly, the high cycle fatigue life is increased by several orders of magnitude in number of load cycles. The increased inter fiber fracture strength could be correlated to the improved fatigue behavior, as final failure in high cycle fatigue is strongly correlated to matrix cracks (Böger et al. 2010).

In order to study the influence of nanoparticles on the fatigue life of GFRPs, quadraxial GF-NCF were fatigue loaded until final failure with constant stress amplitudes at stress ratios of R = 0.1 (tension–tension), R = −1 (tension–compression) and R = 10 (compression-compression), respectively. The number of load cycles to failure was counted for each specimen and plotted versus the respective maximum stress in order to obtain a SN-curve. The elongation of the tested specimens was monitored using an inductive displacement transducer. According to the fatigue test results shown in Fig. 5.12 for the composites under dynamic tensile load reducing the maximum stress amplitude leads to an increase in fatigue life for both the unmodified reference and the 0.3 wt% MWCNT modified one. However, the presence of MWCNTs significantly improves the fatigue life of the composite under cyclic loading by roughly one order of magnitude in load cycles (Böger et al. 2010).

Figure 5.13 depicts the SN-curves of the fatigue tests at a stress ratio of R = −1. Compared to the tests with a stress ratio of R = 0.1, the fatigue tests at a stress ratio of R = −1 show an even more pronounced effect of MWCNT modification on the fatigue life in the high cycle fatigue regime. Due to a lower slope of the SN-curve of the MWCNT-modified composites, the number of load cycles until final failure is shifted by orders of magnitude. However, the general load level of σ_{max} is significantly lower than the load level in the fatigue tests conducted at R = 0.1, emphasising the strong influence of compression load in each consecutive load cycle (Böger et al. 2010).

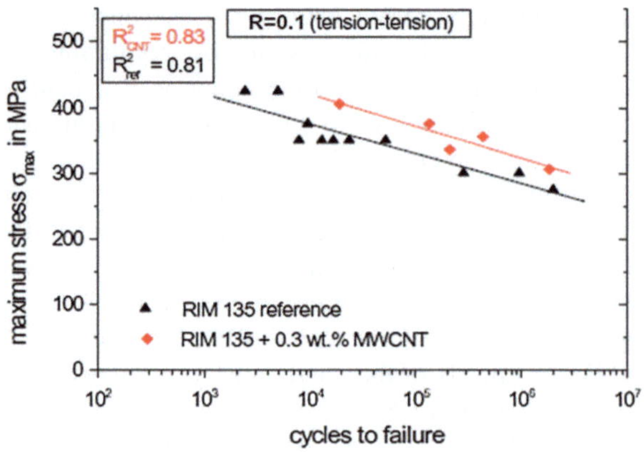

Fig. 5.12 The results reported for stress versus number of load cycles for the composites under dynamic tension-tension load (R = 0.1) (Böger et al. 2010)

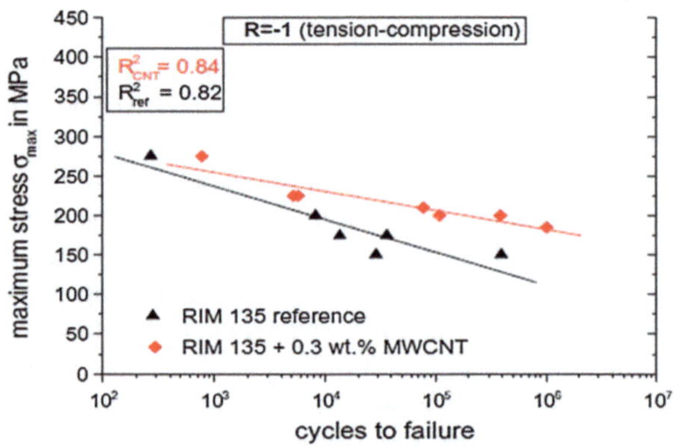

Fig. 5.13 The results reported for stress versus number of load cycles for the composites under dynamic tensile-compression load (R = −1) (Böger et al. 2010)

Figure 5.14 depicts the SN-curves of the fatigue tests at a stress ratio of R = 10. In addition to the SN-curves of the composites with the neat resin and the MWCNT-modified matrix, Fig. 5.14 also includes the data for the composites with the fumed silica-modified matrix. As found for the tests at a stress ratio of R = −1, the SN-curves of both nanoparticle modified composites exhibit a significantly lower slope in comparison to the reference material. For both tests with stress ratios of R = 10 and R = −1, the fatigue life of the nanoparticle modified composites at very high stress levels does not change from the neat matrix composite (Böger et al. 2010).

Shen et al. investigated the effects of incorporating carbon nanotubes (CNT) into polyamide-6 (PA6) on the mechanical performance of woven glass reinforced CNT/PA6 nanocomposite laminates. Results showed that incorporation of up to 2 wt% CNT in CNT/PA6/GF laminates improved the flexural stress of the laminates up to 36 %. The results of tensile tests on CNT/PA6 nanocomposites are summarized in Table 5.3. With increasing CNT content, the tensile modulus and strength increased slightly. The 4 wt% CNT/PA6 nanocomposite

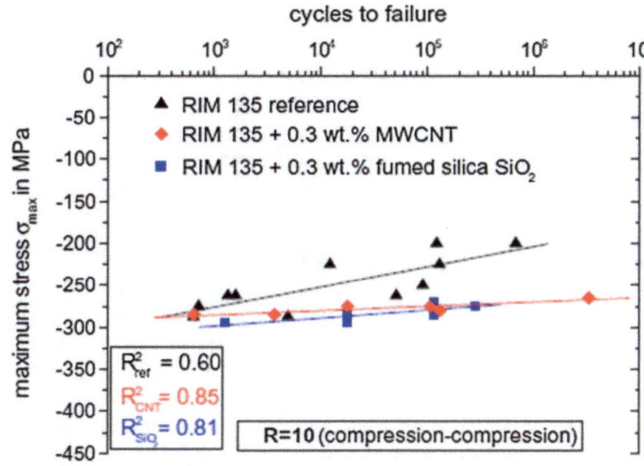

Fig. 5.14 The results reported for stress versus number of load cycles for the composites under dynamic compression–compression load (R = 10) (Böger et al. 2010)

Table 5.3 Tensile test results of CNT/PA6 nanocomposites (Shen et al. 2009)

CNT in matrix (wt%)	1 % Modulus (GPa)	Change (%)	Ultimate tensile strength (MPa)	Change (%)	Elongation at break (%)	Change (%)
0	2.71 ± 0.01	0	74.1 ± 0.17	0	29.5 ± 0.3	0
0.5	2.86 ± 0.02	5.5	80.1 ± 0.15	8.1	30.6 ± 2.3	3.8
1	2.91 ± 0.04	7.4	83.5 ± 0.56	12.7	33.3 ± 1.7	12.9
2	3.00 ± 0.01	10.7	84.3 ± 0.05	13.8	34.4 ± 1.8	16.1
4	3.16 ± 0.03	16.6	87.2 ± 0.12	17.7	29.6 ± 2.7	0.3

led to 17 % increase in the tensile modulus and 18 % increase in the tensile strength without any reduction in elongation compared to the neat PA6 (Shen et al. 2009).

Figure 5.15 shows the percentage increase in flexural modulus and strength of the nanocomposite samples compared to their laminate counterparts. The flexural modulus and strength of CNT/PA6 nanocomposites increased with CNT loading to approximately 20 and 16 %, respectively at 4 wt% CNT, which matched the increases in tensile modulus and strength. However, the increase in flexural performance of the nanocomposite matrix did not transfer according to the properties of the laminates. The flexural properties of the laminates achieved maximum values at a CNT loading of 0.5 or 1 %. The flexural modulus increased by more than 15 % at 0.5 and 1 wt% CNT in PA6 (equivalent to only 0.25 and 0.5 wt% CNT in the laminates), whereas the flexural strength increased to more than 30 % in the same samples. When CNT content increased to more than 1 wt%, the level of increases in flexural modulus and strength reduced even though the absolute values of the samples were still higher than the one without CNT (Shen et al. 2009).

Shekar et al. evaluated interlaminar shear behavior of carbon nanotube and carbon fiber reinforced, epoxy matrix hybrid composite. Short beam shear (SBS) tests were conducted to characterize the influence of fiber orientation in the interlaminar shear strength (ILSS). Their study revealed the importance of fiber alignment

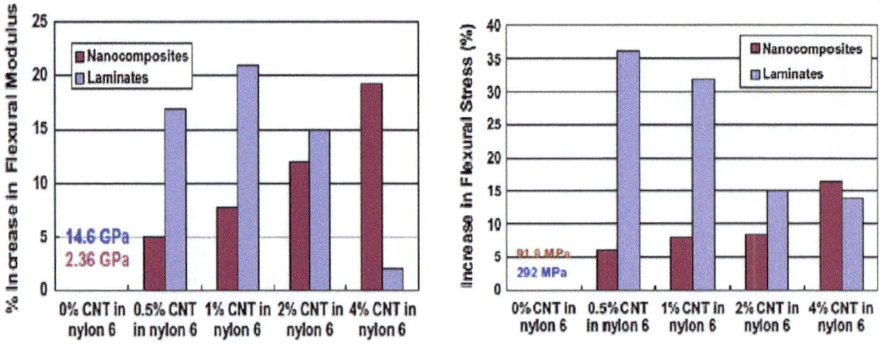

Fig. 5.15 The values of percentage increase in flexural modulus and flexural stress of nanocomposites compared to their laminate counterparts with various amounts of CNT loadings in PA6 (Shen et al. 2009)

and its effects on ILSS properties and nature of deformation under shear loading conditions. Results show that the interlaminar shear strength of the hybrid composite is significantly higher (10 times or nearly 1000 %) in longitudinal orientation as compared to the same in transverse orientation. Based on the results obtained from this study, the hybrid composite material should be designed in a way that the major stress axis is in the longitudinal orientation and should avoid loading in the transverse orientation as much as possible. If the design is based on lowest ILSS values, the true potential of hybrid composite cannot be achieved. This is because the transverse ILSS is an order of magnitude lower. Hence, the hybrid composite studied by Shekar et al. exhibited high anisotropic ILSS properties (Shekar et al. 2014).

5.3.2 Effects of Fiber Modification

Warrier and his co-researchers integrated carbon nanotubes in glass fibers epoxy composites by either including CNTs in the fiber sizing formulation, in the matrix, or both. They studied the effects of such controlled placement of CNTs on the mode I interlaminar fracture toughness of the composites. the presence of CNTs in the sizing resulted in an increased resistance of crack initiation fracture toughness by +10 %, but a lowered crack propagation toughness of −53 %. Similar trends were observed for both instances when CNTs were introduced only in the matrix and in combination of both matrix and sizing (Warrier et al. 2010).

Figures 5.16 and 5.17 show the influence of the presence of CNTs on the behavior of composite during the Mode I interlaminar fracture toughness test (DCB-test): the typical load–displacement diagram and R-curve for fracture toughness for various composites, respectively. Note that the pre-crack length is different for different samples. Initially all the curves show an increase in load,

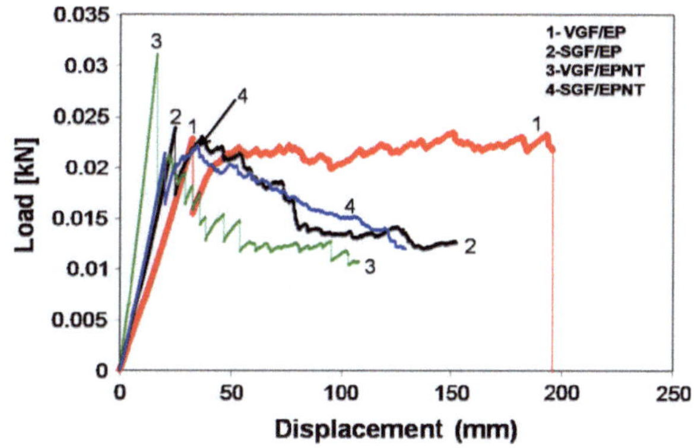

Fig. 5.16 The curves of load–displacement for various composite systems obtained during interlaminar fracture toughness tests (Warrier et al. 2010)

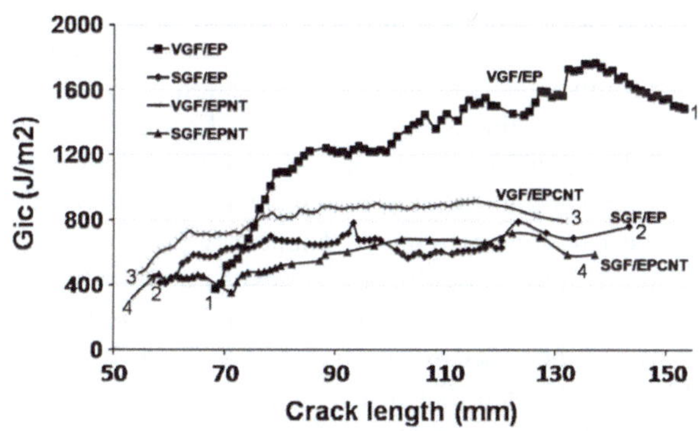

Fig. 5.17 The values of interlaminar fracture toughness (*R-curves*) for various composite systems (Warrier et al. 2010)

but with different slopes, which is due to the different initial crack length and the beam stiffness. After the initial increase in load, there is a large decrease marking the initiation of crack growth. This is again followed by a certain increase in load before it again starts decreasing for all the composite systems studied here, with some noticeable differences in crack propagation pattern (Warrier et al. 2010).

For instance, the VGF/EP load–displacement curve (i.e. curve 1 in Fig. 5.16), after the onset of crack growth and initial decrease, shows a consistent increase and maintains a certain level of load with increased displacement. This indicates a smooth crack growth, which results in a consistent resistance offered by cross-bridging of the VGFs and homogeneity of the matrix micro-structure (Warrier et al. 2010).

Kim et al. processed Hierarchical +1 composites consisting of carbon fibers with carbon nanotubes grown onto them and an epoxy matrix and evaluated the mode I fracture toughness of these composites. Results showed that the mode I fracture toughness of the initial batches of the hierarchical composites was lower than that of the baseline samples without CNTs. Hence, efforts to increase the adhesion between carbon fibers and CNTs were made, resulting in enhanced adhesion. The initial batches of hierarchical composites did not show any improvement

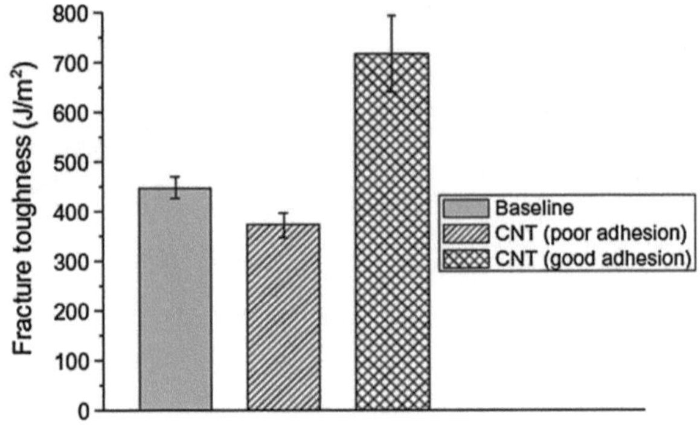

Fig. 5.18 The values of mode I fracture toughness obtained for hierarchical composite specimens (Kim et al. 2015)

in the mode 1 fracture toughness relative to the baseline samples. The results are shown in Fig. 5.18. One plausible reason for the low mode I fracture toughness of the composites is that the poor adhesion between the CNTs and CFs decreased the effective bonding area between the CFs and epoxy matrix (Kim et al. 2015).

After they observed poor fracture toughness for the initial batches, it was concluded that the adhesion between CNTs and CFs is critical for achieving enhanced mode I fracture toughness. Then, they processed the adhesion-enhanced CNT–CF samples into hierarchical composites and evaluated their mode I fracture toughness. The test results showed that the fracture toughness of the adhesion-enhanced samples increased to become 51 % more than that of the baseline samples and 89 % more than that of the hierarchical composites with poor adhesion between CNTs and CFs, as shown in Fig. 5.18 (Kim et al. 2015).

Wang and Zhang studied PopTube technology to use this method to manufacture multiscale, multifunctional structure composites with superior mechanical performance and durability. The PopTube technology is a novel approach to fast grown CNTs on reinforcing fibers in large volume. In this technique, only a single chemical is used to provide both the carbon source and catalyst of CNT growth. Microwave is used to directly heat carbon fibers/fabrics to provide fast and energy-efficient heating. This technique takes only 15–30 s to grow CNTs under the microwave irradiation at room temperature in the air, with no need of any inert gas protection, and additional feed stock gases, typically required in chemical vapor deposition (CVD) approach. Experimental studies revealed that the interlaminar fracture toughness of composite laminates could be significantly enhanced by this technology. After growing CNTs on fabrics using microwave heating, the average interlaminar mode I fracture toughness was increased by 44 % (Wang and Zhang 2013).

Garcia et al. Fabricated a hybrid laminate with aligned carbon nanotubes grown In Situ. They applied a short beam shear test (essentially a 3-point bending test with a short span) to assess any differences in the baseline and CNT hybrid laminates. The test was performed on 4-ply laminates at a constant crosshead speed, and the fracture load and mode were used to determine the interlaminar shear strength (ILSS) of the laminates. Results showed that the values of interlaminar

Table 5.4 The results obtained for tensile strength and stiffness for f-XD-CNT reinforce and neat carbon fiber fabric epoxy composite laminates (Davis et al. 2010)

wt% f-XD-CNT of fabric piece	Ultimate tensile strength UTS (MPa)			Tensile modulus E (GPa)		
	Average	Deviation (%)	% Increase	Average	Deviation (%)	% Increase
0.0	681	±3	0	62	±5	0
0.2	765	–	12	66	–	11
0.3	774	−6, +3	14	76	±5	23
0.5	801	±1	18	77	±5	24

shear strength obtained for the baseline and hybrid laminates were 20.1 ± 0.9 and 33.8 ± 1.1 MPa, respectively. Therefore, the hybrid laminate exhibited an improvement of 69 % with respect to the unreinforced laminate (Garcia et al. 2008a).

Davis et al. studied carbon fiber reinforced epoxy composite laminates, with strategically incorporated fluorine functionalized carbon nanotubes (f-CNTs) at 0.2, 0.3 and 0.5 weight percent (wt%) for improvements in tensile strength and stiffness and durability under both tension–tension (R = +0.1) and tension–compression (R = −0.1) cyclic loadings. The results were then compared with those of the neat (0.0 wt% CNTs) composite laminate material. To develop the nanocomposite laminates, they used a spraying technology to deposit nanotubes on both sides of each four-harness satin weave carbon fiber fabric piece for the 12 ply laminate lay-up. Then, they fabricated the carbon fiber reinforced epoxy laminates using a heated vacuum assisted resin transfer molding (H-VARTM) method followed by a 2 soak curing cycle. As a consequence, significant improvements in the mechanical properties of tensile strength, stiffness and resistance to failure due to cyclic loadings resulted for this carbon fiber reinforced epoxy composite laminate. The tensile test results are given in Table 5.4 (Davis et al. 2010).

For each carbon fiber epoxy laminate material at the prescribed weight percentage (wt%) CNT, the Table presents the average measured ultimate tensile strength (UTS) and modulus (E). Failure in all cases was via tensile fiber rupture. The f-XD-CNTs enhanced the fiber–matrix interfacial strength and the mechanical toughness of the surrounding matrix due to fiber-epoxy–f-CNT interfacial covalent bonding, which accounts for these improvements in laminate strength and stiffness (Davis et al. 2010).

Sharma and Lakkad studied the effect of CNT growth on carbon fibers on the tensile strength of multi-scale composites. The carbon fiber substrate was coated by carbon nanotubes and nanofibers using catalytic decomposition of acetylene precursor and thermal chemical vapor deposition process. These coated carbon fibers were used as reinforcement in epoxy matrix for the fabrication of unidirectional composites. The results of tensile test revealed that both the high vacuum condition and choice of an appropriate catalyst precursor strongly influence the fiber properties thereby affect the resultant properties of the composites made of these surface modified carbon fibers (Sharma and Lakkad 2011).

Table 5.5 The values of tensile strength for composite specimens made of carbon fibers undergone different processing stages in the process of CNT growth on carbon fibers (Sharma and Lakkad 2011)

Type of specimens	Tensile strength (MPa) ± 5
A	360
B	286
C	450
B′	341
C′	581

The composite samples made of carbon fibers with and without CNT growth and undergone different heat treatments were tested for their tensile strength. The composite specimens made of carbon fiber undergone plain heat treatment under low vacuum of 10^{-2} torr are indexed as A, specimens made of carbon fiber undergone heat treatment after acid cleaning and catalyst deposition are indexed as B and specimens made of carbon nanotubes grown carbon fiber are indexed as C. The composites B and C made of carbon fiber undergone heat treatment under high vacuum of the order of 10^{-6} torr are indexed as B′ and C′ respectively. The results obtained from tensile test of these composites are shown in Table 5.5 (Sharma and Lakkad 2011).

As the results show, a clear relative improvement in the strength of composites has been recorded in case of composites C. The tensile strength of the composites C was 57 and 26 % higher than composites B and A respectively. The tensile strength of the composites C′ was 69 and 61 % higher than composites B′ and A respectively. The inset shows the numerical values of tensile strength of different composite specimens. This increase in the tensile strength of composites C′ might be due to stronger fiber/matrix interfacial bonding. They also found that the lack of high vacuum in the pre-growth step of CNTs adversely affects the fiber properties by introducing defects. The fractographic study of the failed composite specimens showed improved interfacial bonding as revealed by less fiber pullout in case of CNTs coated carbon fiber composites (Sharma and Lakkad 2011).

An et al. prepared the multiscale carbon nanotube-hybridized carbon fiber by a newly developed aerosol-assisted chemical vapour deposition. Compared with the original carbon fibers, the fabrication of this hybrid fiber resulted in an almost threefold increase of BET surface area to reach 2.22 m^2/g. Meanwhile, there was a slight degradation of fiber tensile strength within 10 %, while the fiber modulus was not significantly affected, indicating the intrinsic fiber remained undamaged during the CNT growth (An et al. 2012).

In addition, they determined the interfacial shearing strength of a carbon fiber-reinforced polymer composite with carbon nanotube-hybridized carbon fiber and an epoxy matrix from the single fiber pull-out tests of microdroplet composite. Due to an efficient increase of load transfer at the fiber/matrix interfaces, the interracial shear strength of composite reinforced by carbon nanotube-hybridized carbon fiber was almost 94 % higher than that of one reinforced by the original

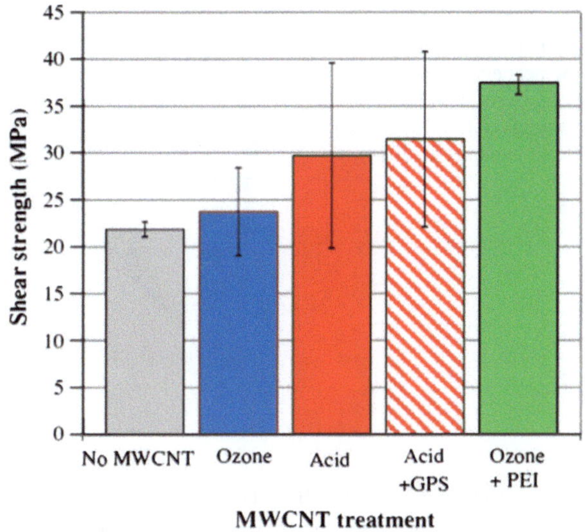

Fig. 5.19 The values of shear strength measured for treated carbon/epoxy composites (An et al. 2012)

carbon fiber. The enhancement of IFSS was induced by the increased surface area, which can improve touching epoxy matrix region surrounded the carbon fiber, thus providing a stronger interfacial bonding between the CNTs, carbon fiber and matrix (An et al. 2012).

An et al. deposited carbon nanotubes (CNTs) onto carbon-fiber fabric using electrophoretic deposition (EPD) prior to the infusion of epoxy resin for the production of carbon/epoxy composites. They functionalized CNTs by using ozone treatment for oxidation, followed by chemical reaction with polyethyleneimine. Their test results showed significant increases in the shear strength and fracture toughness of the carbon/epoxy composites with the CNT treatment relative to composites without treatment. Figure 5.19 shows the influence of the electrophoretically-deposited MWCNTs on the carbon/epoxy in-plane shear strength (An et al. 2012).

The ozone-treated MWCNTs do not appear to make any significant difference to the baseline shear strength, whereas the ozone-PEI treated MWCNTs show almost 70 % increase in the baseline strength (An et al. 2012).

Rahmanian et al. fabricated a high performance multiscale composite by incorporating carbon nanotubes (CNT) and short carbon fibers into an epoxy matrix. CNTs were grown on fibers through chemical vapor deposition (CVD) method. They produced CNT grown short carbon fibers (CSCF) to improve the stress transfer between epoxy and carbon fibers. The studies on multiscale composites revealed significant improvement in elastic strength as well as impact resistance in comparison to the CNT–epoxy or CSCF–epoxy composites. In addition, they found an optimum content of CNT which provided the maximum stiffness and strength. Figure 5.20 illustrates typical stress–strain responses of the neat epoxy and three nanocomposites with the reinforcement content of 0.2 % CNT, 1 % CSCF–0.3 % CNT and 3 % CSCF–0.3 % CNT (Rahmanian et al. 2014).

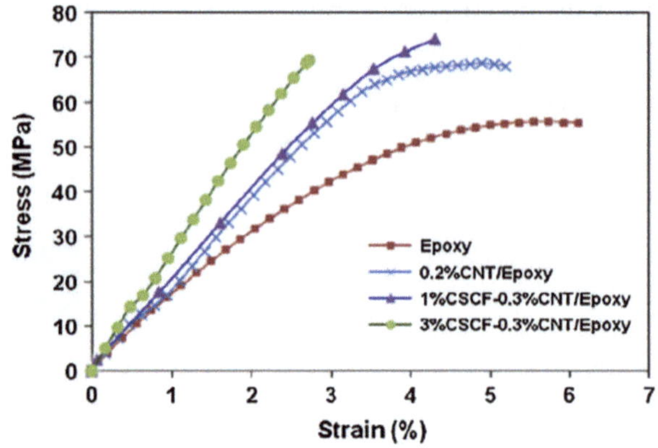

Fig. 5.20 Sample stress–strain curves obtained for the neat epoxy and the composites with different contents of CNT and CSCF (Rahmanian et al. 2014)

As Fig. 5.20 shows, the incorporation of 0.2 wt% CNT to the epoxy matrix did not show notable enhancement of elastic modulus and only modest improvement of tensile strength was achieved. In contrast, incorporation of 1 wt% CSCF increased both modulus and ultimate tensile strength moderately with improvements higher than CNT–epoxy. However, the fracture strain declined significantly in the case of CSCF–epoxy while CNT reinforcement did not change it noticeably. Incorporation of 1 wt% CSCF with CNT–epoxy led to fabrication of multiscale composites which show a remarkable increase of both modulus and strength compared to the baseline epoxy. Modulus and strength of 0.2 % CNT–1 % CSCF composite has been raised to 2.37 GPa and 75.49 MPa respectively in comparison with 1.75 GPa and 55 MPa for the baseline epoxy (Rahmanian et al. 2014).

The maximum improvements were achieved for 0.3 % CNT–1 % CSCF composite with enhancement of 38.8 % in elastic modulus as well as 37.3 % in strength. Additional portions of CNT to 0.4 and 0.5 wt% lowered tensile properties gradually which can be related to the dispersion morphology of CNT in the epoxy matrix. By increasing the CNT content, the probability of creating CNT aggregation is intensified which results in strength deterioration (Rahmanian et al. 2014).

The impact strength of 0.2 % CNT–epoxy reached 1.67 kJ/m^2, whereas the epoxy filled by 1 % CSCF was strengthened in impact resistance up to 1.95 kJ/m^2 equivalent to 38 % improvement. The composite with multiscale reinforcements of 0.2 % CNT–1 % CSCF showed a slightly better enhancement than 1 % CSCF–epoxy composite. Incorporation of additional CNT up to 0.5 wt% fraction did not change the results noticeably indicating that the microfibers play the main role in the composite capacity to absorb impact energy. Unlike the CNT, increasing the CSCF content of 0.2 % CNT–epoxy to 3 wt% led to the maximum impact strength of this study with 56 % improvement for 0.2 % CNT–3 % CSCF–epoxy (Rahmanian et al. 2014).

5.3.3 *Effects of Prepreg Modification*

Siddiqui et al. produced prepregs of carbon fiber-reinforced polymer (CFRP) composites containing carbon nanotubes (CNTs) based on a solventless prepregging process. They studied Rheological properties and the results confirmed that high-speed shear mixing and functionalisation of CNT are effective in lowering the viscosity of the CNT-epoxy nanocomposites (Siddiqui et al. 2011).

Because viscosity is the single most critical parameter for processing of prepregs, a series of experiments was conducted to identify optimal processing conditions with improved CNT dispersion while maintaining the viscosity within the limits. Several important parameters, such as the effects of CNT content, mixing speed, type of epoxy and CNT functionalisation, were evaluated. The potential adverse effects of high CNT contents on the viscosity of resin and the thickness of prepreg were reduced by increasing the resin temperature, fiber tension and the exit die gap of the prepregger. These prepregging parameters were optimized through extensive preliminary studies (Siddiqui et al. 2011).

Ma et al. developed a novel method to realize the situ accumulation of carbon nanofibers (CNFs) in the carbon fiber reinforced polymer composites (CFRPs) to construct the multi-scale reinforcement for improving the interlaminar properties. In this method, the prepreg was sealed by the nanomicroporous nylon membrane, and the excess resin was extracted from the prepreg by the vacuum-assisted method. The use of nylon membrane resulted in effective CNF accumulation (Ma et al. 2015).

They characterized the mode-II interlaminar fracture toughness of the laminate panels by the end-notched flexure test. The results are illustrated in Fig. 5.21. It is seen that there is a gap between the two diagrams as CNFs increase from 1 to 4 wt%, which means that the membrane-assisted method is more efficient for improving the ILSS of the laminate panels than the no membrane-assisted method (Ma et al. 2015).

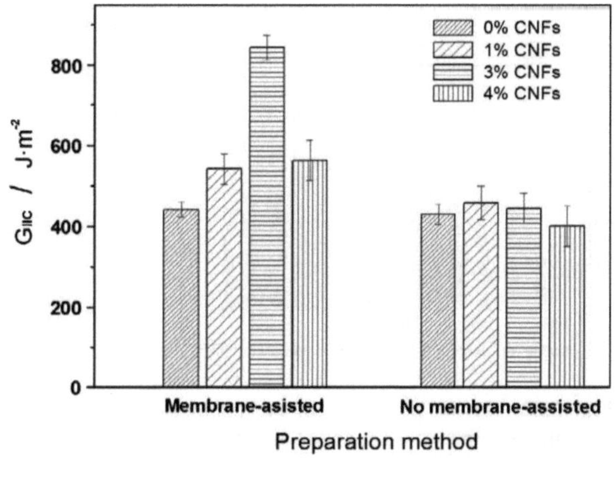

Fig. 5.21 The values of mode-II interlaminar fracture toughness of CFRLC with and without membrane-assisted method (Ma et al. 2015)

Siddiqui et al. fabricated multiscale composites using carbon fiber prepregs containing CNT modified epoxy matrix. They found that the interlaminar shear strength (ILSS) of the composites with 0.5 wt% CNTs increased by 12 % compared to the neat CFRP composites, whereas the torsional shear modulus and strength determined using tubular specimens presented corresponding improvements of 17 and 19.5 %, respectively. The authors suggested that the enhanced interfacial adhesion between the modified matrix and carbon fibers, and the increases in matrix modulus/strength arising from the well-dispersed functionalized CNTs were mainly responsible for the above observations. An increase in CNT content beyond 0.5 wt% had moderately negative effects on these properties, indicating potential agglomeration at a high CNT content (Siddiqui et al. 2013).

Garcia et al. studied a method of joining prepreg composite interfaces with aligned carbon nanotubes. The aligned CNTs were grown at high temperature and transferprinted to prepreg at room temperature, maintaining CNT alignment in the through-thickness direction. The fabrication methods were suggested to be compatible with existing manufacturing processes and to have the potential to enhance the structural and multifunctional properties of advanced composite laminates. Mode I and mode II, hybrid laminates with CNTs displayed increased fracture toughness over the baseline laminates. The results show that the CNT-modified interface increases fracture toughness 2.5X in mode I, and 3X in mode II. Both compliant interlayer and bridging are considered as mechanisms of toughening, with evidence of CNT bridging observed in the fracture micrographs (Garcia et al. 2008b).

Zhang et al. studied mechanical properties of carbon fiber (CF) reinforced epoxy composites enhanced with a small quantity of multi-walled carbon nanotubes (MWCNTs). They dispersed 1 wt% MWCNTs into the epoxy matrix and manufactured the CF prepregs using fiber filament wind process. Then, the hybrid MWCNT/CF/epoxy composites were fabricated by laying up of fiber prepregs. Figures 5.22 and 5.23 show the obtained tensile properties and ILSS of the composites (Zhang et al. 2013).

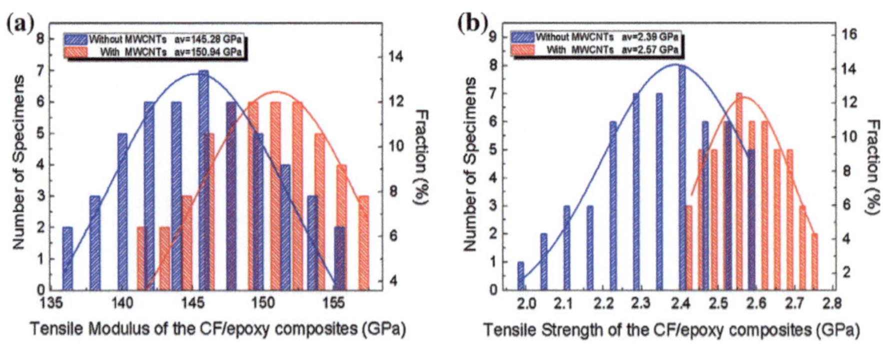

Fig. 5.22 Tensile properties of the composites, **a** Tensile modulus of composites, **b** Tensile strength of composites (Zhang et al. 2013)

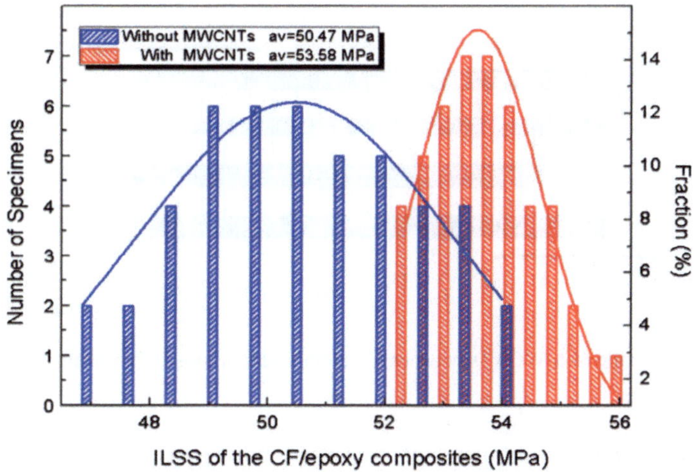

Fig. 5.23 The curves obtained for ILSS of the composites (Zhang et al. 2013)

From Fig. 5.22a, it can be seen that the mean tensile modulus of the nano-modified composites increased, while the dispersity of modulus did not change much. This behavior can be attributed to the following two main effects: First, the modulus of the composites is mainly dominated by the modulus of the CFs rather than the matrix. Hence, the dispersity of the modulus of composites is mainly determined by the dispersity of modulus of the CFs. Second, the dispersion of MWCNTs in the matrix is not perfectly homogenous. The modulus describes the resistance to elastic deformation which is reversible, and is much more sensitive to the defects brought by MWCNT agglomeration (Zhang et al. 2013).

It can be seen from Fig. 5.23 that the ILSS of the composites improved and the dispersity decreased with the introduction of MWCNTs. Generally, interlaminar shear failure involves several fracture mechanisms that occur during delamination, including matrix deformation and fracture, fiber pullout, fiber bridging, and fiber–matrix interface debonding. Matrix and interface enhancing mechanisms for tensile properties also contribute to the increase of ILSS which is a matrix-dominated property. Besides, MWCNTs contribute to the ILSS improvement through another two effects: First, the MWCNTs are randomly dispersed in the matrix. Part of the MWCNTs that oriented perpendicular to the CF cloth layer plane can act as inter-plane bridges and enhance the ILSS effectively, as shown schematically in Fig. 5.24. Second, delamination along the fiber–matrix interface is one of the major failure modes and hence the interface characteristic is critical for the ILSS (Zhang et al. 2013).

Dispersity of mechanical properties of the composites is affected by the inhomogeneity (mainly brought in by defects and residual stress), commonly located at stress concentration area inside the matrix or near the fiber–matrix interface. The MWCNT bridging, pullout and rupture mechanisms discussed above are also

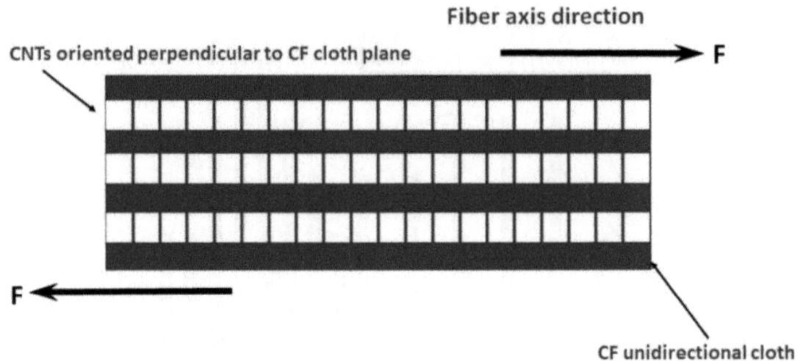

Fig. 5.24 Schematic of MWCNTs between CF cloth layers proposed for enhancing the ILSS

expected to reduce the influence of defects by hindering the initiation and propagation of micro-cracks. In addition, the embedded MWCNTs can reduce the residual stress inside the matrix. Hence, dispersity of the mechanical properties of nano-modified composites can be decreased as a result of the added MWCNTs, as shown by the narrower curves of property distributions for nano-modified composites in Figs. 5.22 and 5.23 (Zhang et al. 2013).

5.3.4 Effects of Secondary Nanoparticles

Just recently, some researchers added a secondary nanoparticle to study its effects on the mechanical properties of CNT-reinforced composites. For instance, Jia et al. studied the multiscale reinforcement and interfacial strengthening on epoxy-based composites by nanoscale complex composed of zero-dimensional silica nanoparticles (SiO_2) and one-dimensional multiwalled carbon nanotubes. Results showed that mechanical properties of epoxy (EP) composites were significantly enhanced by SiO_2–MWCNTs rather than other functionalized MWCNTs, due to synergy reinforcing effect of MWCNTs and SiO_2 as well as enlarged interfacial areas by SiO_2. The chemically bonded nanoscale interfacial area between glass fiber and matrix was generated and bridged by SiO_2–MWCNTs, making glass fiber like a branched reinforcement, resulting in strong interfacial adhesion and effective stress transfer. Figure 5.25 shows the tensile strength and modulus of epoxy reinforced with 0.5 wt% loading of various MWCNTs (Jia et al. 2013).

Tensile strength and modulus of epoxy with functionalized MWCNTs were higher than those of the pristine epoxy and epoxy with r-MWCNTs, since functional groups bonded onto the surface of MWCNTs strengthened the interfacial adhesion between MWCNTs and surrounding epoxy matrix, resulting in an effective stress transfer between their interface and resin matrix. This result confirmed the ameliorating effect of surface functionalization of CNTs on the improvement of mechanical properties of the composites. Among various

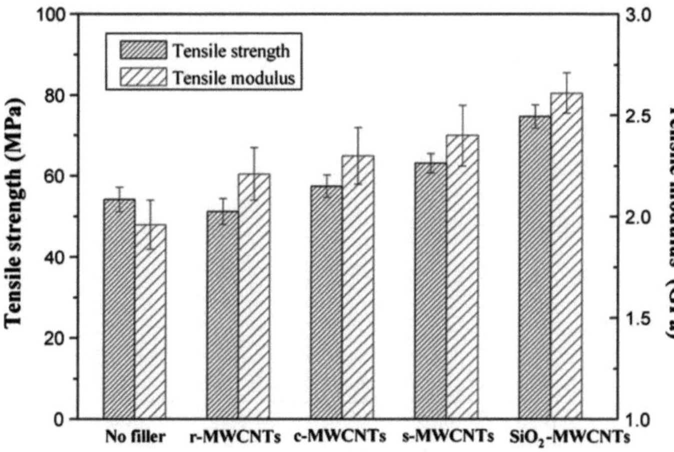

Fig. 5.25 The values of tensile strength and tensile modulus of epoxy reinforced with 0.5 wt% loading of various MWCNTs, respectively (Jia et al. 2013)

functionalized MWCNTs, the SiO_2–MWCNTs showed the highest ameliorating effect. Compared to tensile strength and modulus values of 54.2 ± 3.0 MPa and 1.96 ± 0.12 GPa for the pristine epoxy matrix, respectively, the incorporation of the SiO_2–MWCNTs led to an obvious improvement of tensile strength and modulus by 37.8 and 33.2 %, respectively. This significant improvement was attributed to the synergy reinforcing effect of one-dimensional MWCNTs and zero-dimensional silica nanoparticles as well as the larger interfacial areas between SiO_2–MWCNTs and epoxy of CNT/phenolic composites and pointed out that high loading of the reinforcements caused insufficient impregnation of the resin within the CNTs network (Jia et al. 2013).

Figure 5.26 shows mechanical properties of GF/EP composites with various loadings of SiO_2–MWCNTs. Mechanical properties of GF/SiO_2–MWCNT/EP composites were apparently higher than those of SiO_2–MWCNT/EP composites, showing the reinforcing effect of glass fibers. Moreover, a significant increase in mechanical properties of GF/SiO_2–MWCNT/EP composites was observed up to a SiO_2–MWCNT loading of 0.5 wt% as shown in Fig. 5.26, and then the declining happened at higher loadings of SiO_2–MWCNTs. The declining was also due to gradual aggregation of SiO_2–MWCNTs at higher loadings. At SiO_2–MWCNTs loading of 0.5 wt%, tensile strength, elongation at break and Young's modulus of the composites were 902 ± 13 MPa, 2.70 ± 0.25 % and 46.1 ± 1.8 GPa, respectively, achieving the corresponding improvement of 9.3, 7.6, and 8.5 % in comparison with pristine GF/EP composites (0 wt% SiO_2–MWCNTs) (Jia et al. 2013).

The notable improvement of the tensile properties of GF/SiO_2–MWCNT/EP composites was related with chemically crosslinked molecule network formed by the reaction of epoxy matrix with the amino groups on the surfaces of MWCNTs, accompanied with the additional reinforcing by zero-dimensional silica

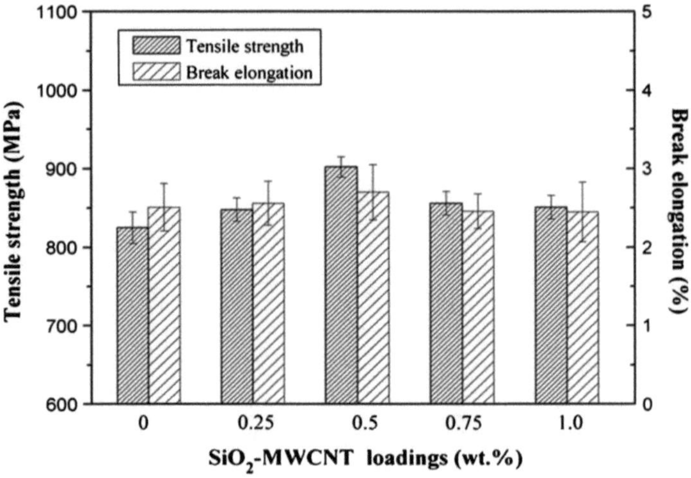

Fig. 5.26 The tensile strength and elongation at break for GF/EP composites with various loadings of SiO_2–MWCNTs (Jia et al. 2013)

nanoparticles surrounding one-dimensional MWCNTs. In this case, it was noted that the SiO_2–MWCNTs acted as three-dimensional branched reinforcement, not the normal one-dimensional cylindrical filler (Jia et al. 2013).

With the aim of fundamentally understanding the role of SiO_2–MWCNTs on the interfacial adhesion between glass fiber and epoxy matrix, ILSS of the composites, which was generally considered as a measure of the adhesive bonding strength, were characterized. Figure 5.27 shows the ILSS of GF/EP composites with various loadings of SiO_2–MWCNTs (Jia et al. 2013).

The results showed that the incorporation of 0.25, 0.5, 0.75 and 1.0 wt% SiO_2–MWCNTs enhanced the ILSS by 10.6, 25.9, 17.3, and 11.7 % with respect to the pristine GF/EP composites, respectively. This revealed that SiO_2–MWCNTs were highly effective in strengthening interfacial load transfer between the glass fiber and the epoxy matrix, since the incorporation of SiO_2–MWCNTs resulted in MWCNT-bridging and chemical interlocking between fiber and matrix and improved the toughness of the resin matrix (Jia et al. 2013).

Li et al. incorporated hybrid fillers comprised of CNTs directly grown on alumina micro-spheres by chemical vapor deposition into epoxy matrix that was then reinforced with woven glass fibers. They used randomly distributed CNTs, reported as ran-CNTs, with 30 nm average diameter and 20 μm average length, and Al_2O_3 micro-spheres with 3–10 μm in diameter for fabricating hierarchical composites. The CNT–Al_2O_3 hybrids with multi-walled CNTs grown on the Al_2O_3 particles were synthesized by CVD (He et al. 2010). The hierarchical composites with 0.5 wt% hybrid loading were observed to exhibit an improvement of 19 and 11 % in flexural modulus and interlaminar shear strength, respectively. These reinforcements are mainly attributed to the improvements of matrix

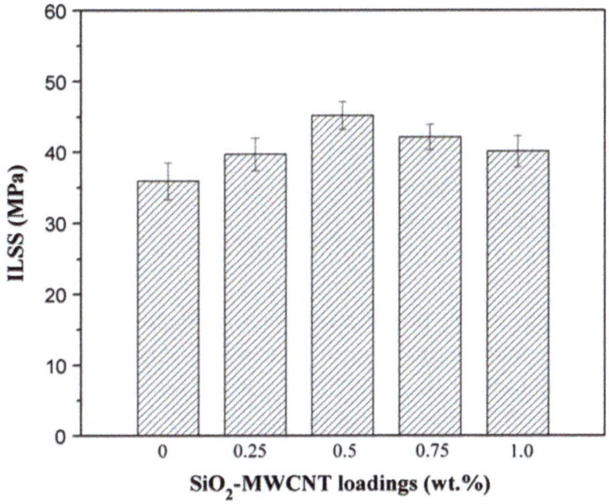

Fig. 5.27 The values of ILSS obtained for GF/EP composites with various percent loadings of SiO_2–MWCNTs (Jia et al. 2013)

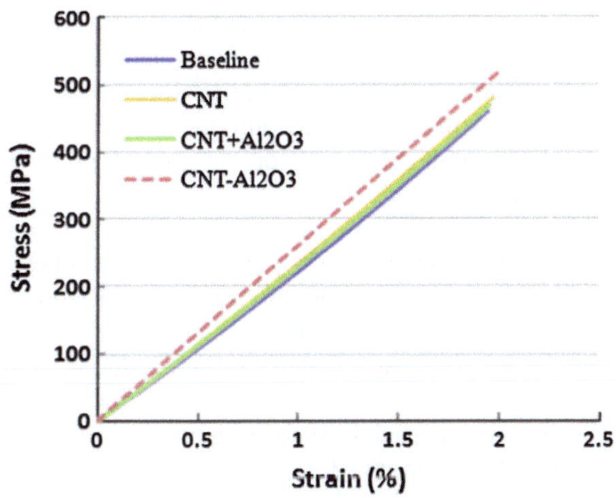

Fig. 5.28 Sample stress–strain curves obtained for flexural properties of the glass fabric/epoxy composites (Li et al. 2014)

properties resulted from the good dispersion of hybrids and their hindering effect on the formation and development of matrix cracks. Figure 5.28 shows the typical stress–strain curves of the glass fabric reinforced neat and modified epoxy matrix with the same concentration (0.5 wt%) of ran-CNTs, ran-CNT + Al_2O_3 and CNT–Al_2O_3 hybrids under three-point bending test (Li et al. 2014).

As Fig. 5.28 shows, the flexural modulus and strength of glass fabric/epoxy/CNT–Al_2O_3 composites evidently increases, while the other two types of composites show somewhat lower increment. Given the introduction of CNT–Al_2O_3, the flexural modulus is up to 26 GPa, corresponding to 19 % increase over the reference samples, and their flexural strength is increased by 12 %. It indicates

Table 5.6 The values of ILSS obtained for the glass fabric/epoxy composites with and without filler modification (Li et al. 2014)

Sample	ILSS (MPa)	Improvement (%)
Glass fabric/epoxy	26.3 ± 0.8	–
Glass fabric/epoxy/ran-CNT	28.0 ± 0.7	6.5
Glass fabric/epoxy/ran-CNT + Al_2O_3	27.5 ± 0.9	4.6
Glass fabric/epoxy/CNT–Al_2O_3	29.1 ± 0.8	10.7

that the fillers provide reinforcing mechanism via matrix strengthening, which is more evident in the case of CNT–Al_2O_3 hybrids. Li et al. also investigated interlaminar shear strength (ILSS) of reference sample and multi-scale composites containing 0.5 wt% of ran-CNTs, ran-CNT + Al_2O_3 and CNT–Al_2O_3 under three-point short beam shear testing and the results are presented in Table 5.6 (Li et al. 2014).

Compared to the reference samples, the addition of ran-CNT and ran-CNT + Al_2O_3 slightly increases the ILSS values by 6.5 and 4.6 %, respectively. Meanwhile, the CNT–Al_2O_3 further enhances the ILSS values to 29.1 MPa (10.7 % increase). This can be attributed to the good dispersion of the hybrids in the matrix, leading to higher modulus and lower viscosity, as well as the strong interfacial interactions between fillers and matrix, where the hybrids can serve as bridges interconnecting neighboring fibers to each other. Although further optimizations are still required, the introduction of multiscale hybrids endows the fibrous composites with enhanced properties, which provides greater potential for applications (Li et al. 2014).

Barbaz-I studied the improvements in the tensile properties of carbon/epoxy composite by adding multi-wall carbon nanotubes and nano-silica. The composite specimens were prepared by vacuum assisted resin infusion molding (VARIM) method. The advantages of this method for fabricating composite structures are high-speed composite manufacturing, reduced fabricating defects due to manufacturing errors compared to hand layup method, and also possibility of adding abrasive nanoparticles, such as nano-silica, without damage to the equipment compared to resin transfer molding technique (RTM). Nanoparticles were distributed and dispersed in the epoxy resin by using ultrasonic and mechanical stirrer devices. Composite specimens with 0.1, 0.5 and 0.9 wt% nano-silica and carbon nanotubes contents, were fabricated. For the composite containing both nanoparticles, the hybrid composites with (0.2 wt% CNT + 0.7 wt% nano-silica), (0.2 wt% CNT + 0.7 wt% nano-silica) and (0.2 wt% CNT + 0.7 wt% nano-silica) content were prepared (Barbaz-I 2014).

With the addition of only 0.5 wt% carbon nanotube, the improvements were 22.3 % in tensile strength and 27.9 % in tensile modulus, and with the addition of 0.5 wt% nano-silica, 15.3 % in tensile strength and 12.8 % in tensile modulus. With increasing the weight percentage of nanoparticles to 0.9 wt%, the improvement in the tensile properties decreased. As scanning electron microscope images of fracture surfaces in this study showed, the declining was due to aggregation of MWCNTs and nano-silica at higher loadings. The best improvement in tensile properties was obtained for the case of hybrid composite containing (0.45 wt% N–SiO_2 + 0.45 wt% CNT). Results of this case showed that, tensile strength 25.9 % and tensile modulus 31 % were improved, while in the same weight percent (total weight percent of both nanoparticles) of single nanoparticle content, the mechanical properties were decreased (Barbaz-I 2014).

5.4 Tribological Properties

The addition of nanoparticles like CNT to the matrix of polymer-based composites can affect the tribological behavior of composites. Guignier et al. studied the tribological properties of CNTs grafted on carbon fibers by the flame method. The aim of their study was to determine whether CNTs on fibers suffer damage under stress, similar to those applied during composite manufacturing, which can damage composite properties, particularly fiber/matrix adhesion. For this purpose, they performed reciprocating friction tests to examine the resistance of CNTs and to highlight a wear mechanism. The results showed that the presence of CNTs increased the coefficient of friction (COF) in the first friction cycles and then decreased it to close to the COF of the fiber without CNTs. In this study, the normal load was applied by means of dead weights. The friction force was measured using a 3 N strain gauge sensor. The simulator is motor-controlled, whereby a sinusoidal movement could be created. In addition, the contact during friction can be observed laterally by an image acquisition system. Figure 5.29 presents the evolution of the coefficient of friction relative to the number of reciprocating cycles for CNT-grafted fibers (Guignier et al. 2015).

The results in Fig. 2.29 show that the coefficient of friction at the first cycle of friction is higher for CNT-grafted fibers (0.27) than for virgin fibers (0.24), indicating the CNT influence on the coefficient of friction. A decrease appears for both samples, more rapidly for virgin fibers, and then the coefficient of friction stabilised at around 0.23. This can partly be attributed to rearrangement of the textile structure. Indeed, the loose interlacing of the warp and weft allows the mobility of the fibers in the structure. This mobility exists for grafted fibers because the CNTs cover the fibers and do not stick the fibers together (Guignier et al. 2015).

The COF decreases more slowly for the CNT-grafted fabric than for the virgin fabric (Fig. 5.29). Two mechanisms might be suggested for this behavior: (i) a mechanical origin due to the presence of CNTs acting like bristles, increasing the friction with the slider and/or between the fibers because of entanglement between CNTs, or (ii) a physicochemical mechanism due to greater adhesion between the fibers and the slider because of a larger specific surface area of the fibers with CNTs (Guignier et al. 2015).

They also studied influence of the normal load on the friction coefficient. Figure 5.30 presents the evolution of coefficient of friction under different normal loads.

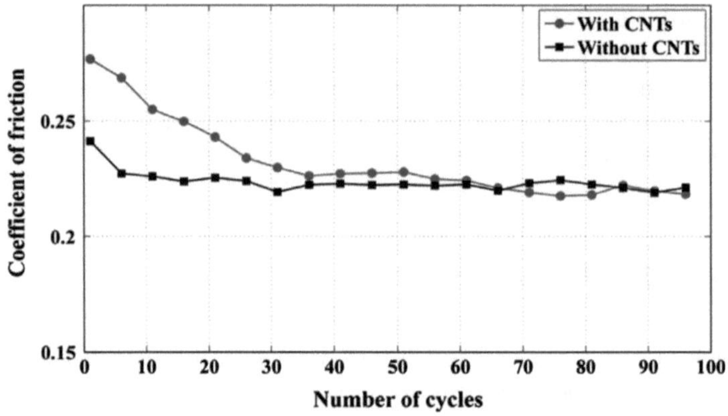

Fig. 5.29 The variations of friction coefficient for carbon fibers grafted with CNTs (*circles*) and carbon fibers without CNTs (*squares*) as a function of the number of cycles under a normal load of 2 N (Guignier et al. 2015)

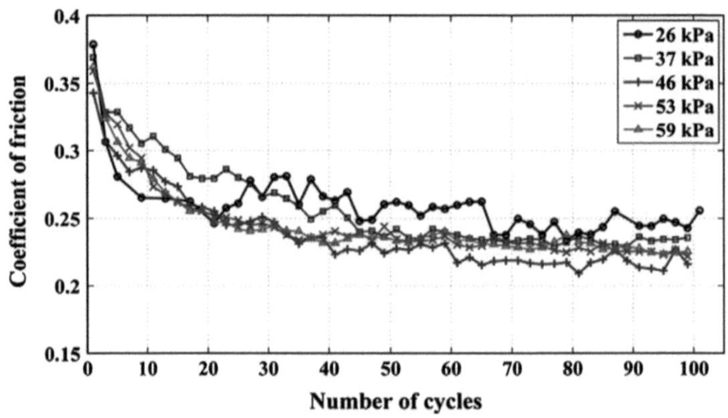

Fig. 5.30 The variations of friction coefficient for carbon fibers grafted with CNTs under different mean pressures: 26 kPa (*circles*), 37 kPa (*squares*), 46 kPa (*plus symbols*), 53 kPa (*crosses*) and 59 kPa (*triangles*) (Guignier et al. 2015)

The results confirm the decrease in the coefficient of friction during the first cycles of friction regardless of the applied normal load. Over 50 cycles, the coefficient of friction appeared to balance out at a value between 0.22 and 0.25 (Guignier et al. 2015).

A study of the wear mechanism revealed the behavior of CNTs as a transfer film, which explained the decrease in the coefficient of friction because a transfer film plays the role of a lubricant and diminishes friction resistance. Moreover, CNTs have good adhesion with the fiber, because they endure friction; the transfer film is only removed after 50 cycles of friction, which is more important than the real strain applied to the fabric during composite manufacturing. After removing

Table 5.7 The fiction coefficients obtained for three composites, Load is 49.03 N and Speed is 0.18 m/s (Kim et al. 2011)

Specimen	Coefficient of friction
Carbon/epoxy	0.3–0.6
Acid-treated carbon/CNT/epoxy	0.2–0.45
Silane-treated carbon/CNT/epoxy	0.15–0.4

the transfer film, the friction behavior is close to that of the original carbon fiber fabric (Guignier et al. 2015).

Kim et al. investigated the effect of carbon nanotube (CNT) modifications on the wear behaviors of multiscale carbon/CNT/epoxy composites. They fabricated Carbon/epoxy woven composites and two types of multiscale carbon/CNT/epoxy composites by incorporating woven-type carbon fibers into epoxy matrices modified with 2 wt% acid-treated and silane-treated multi-walled carbon nanotubes (MWCNTs). Ball-on-disk wear tests were performed on the three composites. The results showed that the wear properties of carbon/epoxy composites were improved by the addition of CNTs. In addition, the wear properties of silane-treated samples were superior to those of acid-treated samples. Table 5.7 shows the change in friction coefficient as a function of wear distance for each of the three composites (Kim et al. 2011).

As shown in the table, a significant difference is observed among the friction coefficient values of the carbon/epoxy composite and multiscale carbon/CNT/epoxy composites. The friction coefficient of the carbon/epoxy composite was in the range of 0.3–0.6. In contrast, it fell in the range of 0.15–0.45 for the multiscale carbon/CNT/epoxy composites. It is notable that the surface modification of CNTs produces a lower friction coefficient for the multiscale carbon/CNT/epoxy composites than the carbon/epoxy composite. This result indicates that the modified CNTs improve dispersion and facilitate stronger interfacial bonding between the CNTs and the epoxy matrix (Kim et al. 2011).

Figure 5.31 shows the maximum mean depth profile and the specific wear rate, respectively, for the three composites. The mean depth profile was obtained according to several points taken from the cross-section in the worn area. The maximum depth profile of the carbon/epoxy composite was 294 lm, while the maximum depth profiles of the acid-treated and silane-modified carbon/CNT/epoxy composites were 188 and 164 lm, respectively. The specific wear rate of each composite was determined based on the sectional shape of the wear track. As shown in Fig. 5.30, the multiscale carbon/CNT/epoxy composites exhibited better wear resistance behavior than the carbon/epoxy composite. Specifically, the wear volume loss of the silane-treated specimen was approximately 60 % lower than that of the carbon/epoxy specimen. The enhancement of the wear properties of silane-treated carbon/CNT/epoxy multiscale composites was due to the improved dispersibility and interfacial interaction between the silane-modified MWCNTs and the epoxy in the carbon/epoxy composites (Kim et al. 2011).

In order to clarify the effects of carbon fiber size on the properties of carbon fiber/high strength glass fabric (HSGF)/phenolic laminate, Liu et al. fabricated

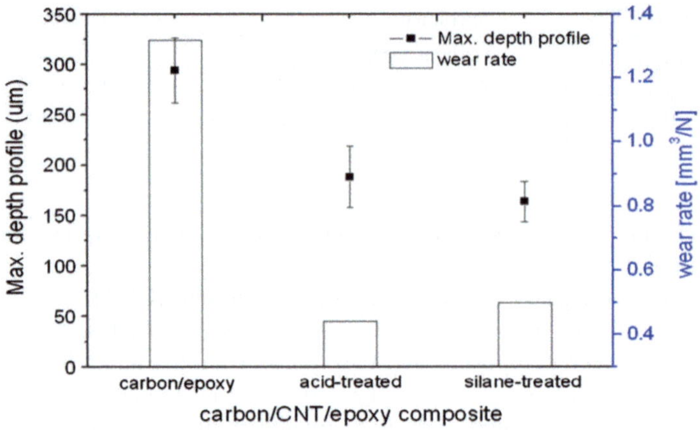

Fig. 5.31 The values of maximum depth profile and wear loss obtained for three composites: carbon/epoxy and multiscale carbon/CNT/epoxy composites (Kim et al. 2011)

two kinds of laminates modified by nano-sized carbon fibers (CNFs) and micro-sized carbon fibers (CMFs), respectively. They investigated tribological properties of HSGF/phenolic laminates modified by CNFs and CMFs in water environment. Results showed that CMFs were more effective than CNFs in improving the wear resistance of HSGF/phenolic laminate in water (Liu et al. 2015). The friction coefficient curves and the specific wear rates of CNFs/HSGF/phenolic and CMFs/HSGF/phenolic laminates are shown in Figs. 5.32 and 5.33.

With the increase of CNF content, the friction coefficient of the laminate gradually decreases. Especially when the content reaches 4.0 %, the friction coefficient is quite stable and lowered to 0.1. This indicates that CNFs play a fundamental role in reducing and stabilizing the friction coefficient of HSGF/phenolic laminate. As for CMFs/HSGF/phenolic laminates, although the friction coefficients

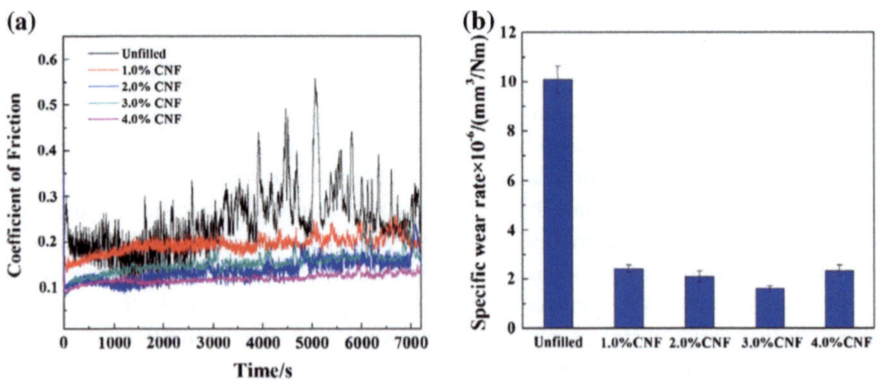

Fig. 5.32 Experimental results for friction coefficient curves (**a**) and specific wear rates (**b**) of CNFs/HSGF/phenolic laminate (Liu et al. 2015)

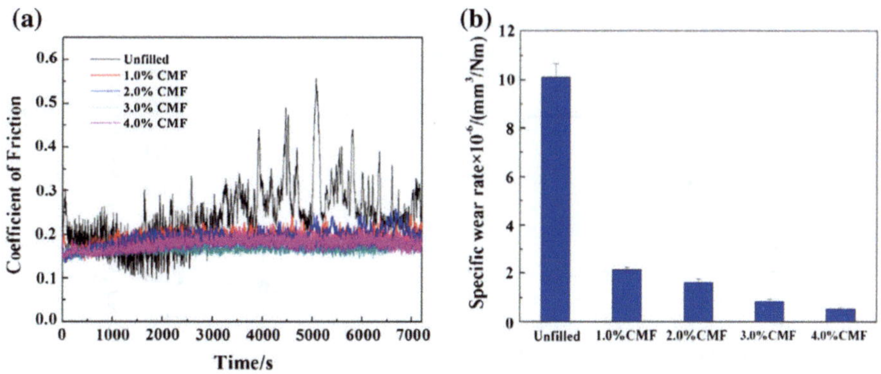

Fig. 5.33 Experimental results for friction coefficient curves (**a**) and specific wear rates (**b**) of CMFs/HSGF/phenolic laminate (Liu et al. 2015)

are lowered and stabilized compared to HSGF/phenolic laminate, the curves are still fluctuating with the value of friction coefficient above 0.15. That is, CNFs are more effective than CMFs in improving friction-reducing property of HSGF/phenolic laminate (Liu et al. 2015).

With the addition of CNFs, the wear rate of HSGF/phenolic laminate is reduced by more than 78 %. Specifically, as the content of CNFs increases from 1.0 to 3.0 %, the wear rate of the laminate gradually decreases. However, when the content of CNFs reaches 4.0 %, the wear rate begins to increase instead. For CMFs/HSGF/phenolic laminate, the addition of CMFs gives rise to a larger reduction of the wear rate of HSGF/phenolic laminate than CNFs. With the increase of CMFs content, the wear rate of the laminate decreases. Especially when the content reaches 4.0 %, the wear rate is almost two orders of magnitude lower than that of HSGF/phenolic laminate. Thus, CMFs play a more effective role than CNFs in improving the wear resistance of HSGF/phenolic laminate (Liu et al. 2015).

As suggested by Lin et al. the greater improvement on wear resistance of HSGF/phenolic laminate by CMFs than CNFs should be attributed to the higher load-capability of CMFs. During the wear process, due to the better wear resistance of CMFs than the resin matrix, they are gradually protruded from the worn surface of laminate after the resin matrix is firstly worn off. Thus, CMFs will be the main bodies directly sliding against the counterface. In the following sliding process, the protruded CMFs will carry the main load between the sliding surfaces, which greatly restricts the further wear of resin matrix and hence the wear of the laminate. The protruded CMFs from the sliding surface, which will cause abrasive wear, should be responsible for higher friction coefficient of CMFs/HSGF/phenolic laminate than that of CNFs/HSGF/phenolic laminate (Liu et al. 2015).

Zhang et al. studied the enhanced wear properties of hybrid PTFE/cotton fabric with phenolic matrix composites filled with functionalized multi-walled carbon nanotubes. For funcionalising MWCNTs, 100 mg of MWCNTs–CH_2OHwere added into a mixture of 100 ml of ethyl acetate, 1 g maleic anhydride and 10 ml

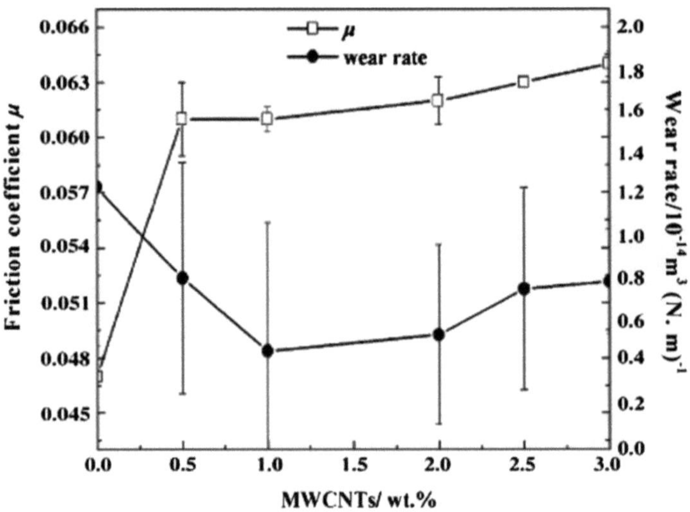

Fig. 5.34 The variations of friction coefficient and wear rate obtained for the hybrid PTFE/cotton fabric composites with the content of MWCNTs under the load of 203.84 N at room temperature (sliding speed: 0.26 m/s) (Zhang et al. 2009)

of perchloric acid. The mixture was sonicated for 40 min and refluxed at 60 °C for 4 h. The product (denoted as MWCNTs-g-MA) was washed with water and dried in vacuum. They performed Pin-on-disk wear tests to investigating wear properties of composites. The pin stays over the disk with two degrees of freedom: a vertical one, for normal load application by direct contact with the disk, and a horizontal one, for friction measurement. The tribological properties of synthesized hybrid PTFE/cotton fabric composites with different contents of MWCNTs were further tested. Figure 5.34 shows the friction coefficient and the wear rate of the hybrid PTFE/cotton fabric composites plot as a function of the content of MWCNTs under the load of 203.84 N at room temperature (Zhang et al. 2009).

As shown in Fig. 5.34, the addition of very small amount of MWCNTs (0.5 wt%) can reduce the wear rate of the hybrid PTFE/cotton fabric composite. A significant reduction of wear rate is observed when the proportion is increased to 1 wt%. If greater proportions of MWCNTs are added, a correlation between the wear rate and the amount of filler is observed. Therefore, the wear rate can be considered to be proportional to the amount of filler. In the sliding process, the surface resin was partially removed and MWCNTs were then exposed to the sliding interface. The exposed MWCNTs had the effect of carrying load and protected the composite from being destroyed, which can be attributed to excellent mechanical properties of MWCNTs. As for the friction coefficient, it varied between 0.061 and 0.064 with different filler proportion. Since the optimal tribological properties were obtained when the MWCNT proportion was 1 wt%, this content was chosen to investigate the effect of load on the friction and wear behaviors of hybrid fabric composite filled with MWCNTs or MWCNTs-g-MA (Zhang et al. 2009).

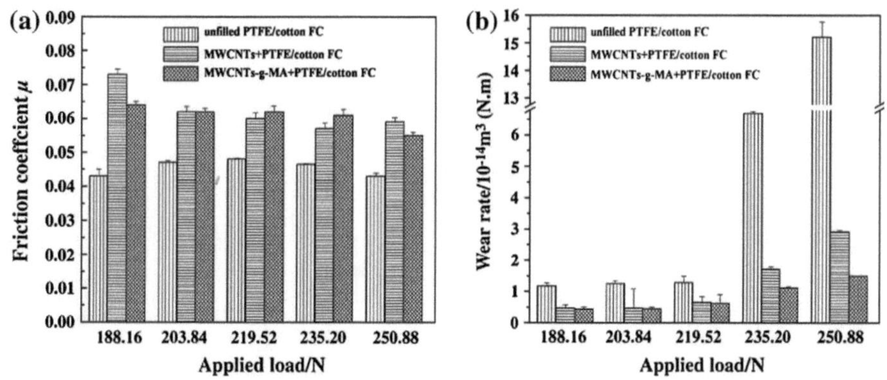

Fig. 5.35 The variations of friction coefficient (**a**) and the wear rate (**b**) obtained for hybrid PTFE/cotton fabric composites unfilled, filled with MWCNTs, MWCNTs-g-MA with increasing applied load at room temperature (Zhang et al. 2009)

They also studied the influence of applied load on the friction coefficient and wear rate. Figure 5.35 shows the variations of the friction coefficient and wear rate of hybrid PTFE/cotton fabric composite unfilled, filled with MWCNTs or MWCNTs-g-MA with increasing applied load.

It can be seen from Fig. 5.35a that the friction coefficient of all the three types of composites did not fluctuate much with the load increasing, which may be caused by the PTFE fiber involved in the composites. Figure 5.35b reveals that under lower applied loads, the wear rate of three types of composites did not differ from each other much. But under higher loads, the wear rate of the unfilled fabric composite increased significantly, which may be caused by the mechanical incompatibility between the hybrid composite and substrate. Fabric composite filled with MWCNTs can decrease the wear rate to a certain extent. However, due to their poor compatibility with matrix, they tended to agglomerate and failed to provide excellent tribological properties as desired. Fabric composite filled with MWCNTs-g-MA exhibited a more favorable anti-wear ability under higher loads. It is considered that the active functional groups grafted to MWCNTs were beneficial to the MWCNTs/matrix miscibility and to the improvement of adhesion strength between MWCNTs/matrix, by chain entanglement and/or chemical bonding between the grafted maleic anhydride and the matrix (Zhang et al. 2009).

Gong et al. investigated the effect of CNTs on tribological properties of C/C composites. They fabricated carbon nanotube (CNT)-doped carbon/carbon (C/C) composites by the chemical vapor infiltration (CVI) method. Ring-on-block-type wear tests were performed to evaluate the frictional properties of CNT-doped C/C composites. Wear and friction tests on ring-on-block-type M-2000 wear tester showed that favorable effects of CNTs on tribological properties were achieved by increasing wear resistance and maintaining stable coefficient of friction under different loads for CNT-doped C/C composites (Gong et al. 2006).

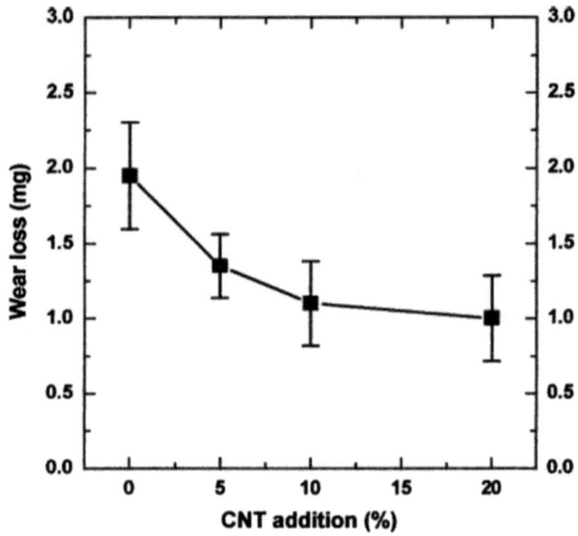

Fig. 5.36 The variations of weight loss in terms of CNT addition after wear test (Lim et al. 2002)

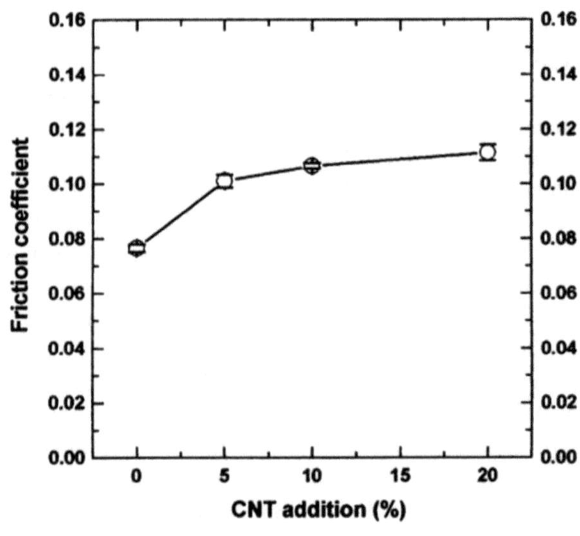

Fig. 5.37 The variations of friction coefficient in terms of CNT addition during wear test (Lim et al. 2002)

Polarized light microscopy, X-ray diffraction, scanning electron microscopy and Raman spectroscopy analyses demonstrated that flake-shaped debris and integrate film on the worn surface could be easily formed for CNT-doped composites, which brought about favorable frictional properties (Gong et al. 2006).

Lim et al. studied the effect of carbon nanotube addition on the tribological behavior of carbon/carbon composites. Ball-on-disc type wear tests were performed to evaluate the tribological properties of the carbon nanotube added carbon composites. The result showed that addition of nanotube had the potential to increase the wear resistance of carbon composites. The variations of wear loss and friction coefficient as a function of CNT addition are shown in Figs. 5.36 and 5.37 (Lim et al. 2002).

Results showed that the friction coefficient slightly increased for higher CNT additions. Weight loss of the 5 % CNT added composite decreased by nearly 30 % as compared to that of the 0 % CNT added composite. As the CNT concentrations increased up to 20 %, the weight losses decreased gradually. The wear resistance of the 20 % CNT added C/C composite was improved by about 100 % as compared to that of the 0 % CNT added C/C composite. The enhanced wear properties were a result of the multi-walled CNTs behavior in reinforcing filler in CNT/carbon composite layer and within the C/C composite (Lim et al. 2002).

References

An, F. et al.: Preparation and characterization of carbon nanotube-hybridized carbon fiber to reinforce epoxy composite. Mater. Des. **33**, 197–202 (2012)

An, Q., Rider, A.N., Thostenson, E.T.: Electrophoretic deposition of carbon nanotubes onto carbon-fiber fabric for production of carbon/epoxy composites with improved mechanical properties. Carbon **50**(11), 4130–4143 (2012)

Arai, M., Noro, Y., Sugimoto, K. i., Endo, M.: Mode I and mode II interlaminar fracture toughness of CFRP laminates toughened by carbon nanofiber interlayer. Compos. Sci. Technol. **68**(2), 516–25 (2008)

Barbaz-I, R.: Experimental determining of the elastic modulus and strength of composites reinforced with two nanoparticles. MSc Thesis, School of Mechanical Engineering Iran University of Science and Technology, Tehran, Iran (2014)

Barber, A.H., Zhao, Q., Wagner, H.D., Baillie, C.A.: Characterization of e-glass polypropylene interfaces using carbon nanotubes as strain sensors. Compos. Sci. Technol. **64**(13–14), 1915–1919 (2004)

Böger, L., Sumfleth, J., Hedemann, H., Schulte, K.: Improvement of fatigue life by incorporation of nanoparticles in glass fibre reinforced epoxy. Compos. A **41**(10), 1419–1424 (2010)

Chen, W., et al.: Basalt fiber–epoxy laminates with functionalized multi-walled carbon nanotubes. Compos. A **40**(8), 1082–1089 (2009)

Davis, D.C., Wilkerson, J.W., Zhu, J., Ayewah, D.O.O.: Improvements in mechanical properties of a carbon fiber epoxy composite using nanotube science and technology. Compos. Struct. **92**(11), 2653–2662 (2010)

Fan, Z., Santare, M.H., Advani, S.G.: Interlaminar shear strength of glass fiber reinforced epoxy composites enhanced with multi-walled carbon nanotubes. Compos. A **39**(3), 540–554 (2008)

Garcia, E.J., Wardle, B.L., Hart, A.J., Yamamoto, N.: Fabrication and multifunctional properties of a hybrid laminate with aligned carbon nanotubes grown In Situ. Composites Science and Technology **68**(9), 2034–2041 (2008a)

Garcia, E., Wardle, B., Hart, A.: Joining prepreg composite interfaces with aligned carbon nanotubes. Compos. A **39**(6), 1065–1070 (2008b)

Godara, A., et al.: Influence of carbon nanotube reinforcement on the processing and the mechanical behaviour of carbon fiber/epoxy composites. Carbon **47**(12), 2914–2923 (2009)

Gong, Q.M., et al.: Tribological properties of carbon nanotube-doped carbon/carbon composites. Tribol. Int. **39**(9), 937–944 (2006)

Green, K.J., Dean, D.R., Vaidya, U.K., Nyairo, E.: Multiscale fiber reinforced composites based on a carbon nanofiber/epoxy nanophased polymer matrix: synthesis, mechanical, and thermomechanical behavior. Compos. A **40**(9), 1470–1475 (2009)

Guignier, C., et al.: Tribological behaviour and wear of carbon nanotubes grafted on carbon fibres. Compos. A **71**, 168–175 (2015)

He, D., Bozlar, M., Genestoux, M., Bai, J.: Diameter- and length-dependent selforganizations of multi-walled carbon nanotubes on spherical alumina microparticles. Carbon **48**(4), 1159–1170 (2010)

Isayev, A.I., Kumar, R., Lewis, T.M.: Ultrasound assisted twin screw extrusion of polymer-nanocomposites containing carbon nanotubes. Polymer **50**(1), 250–260 (2009)

Ismagilov, Z.R., et al.: Development of methods of growing carbon nanofibers on silica glass fiber supports. Catal. Today **102–103**, 85–93 (2005)

Jia, X. et al.: Multiscale reinforcement and interfacial strengthening on epoxy-based composites by silica nanoparticle-multiwalled carbon nanotube complex. Compo. Part A, **48**, 101–109 (2013)

Khan, S.U., Kim, J.K.: Impact and delamination failure of multiscale carbon nanotube-fiber reinforced polymer composites: a review. Int. J. Aeronaut. Space Sci. **12**(2), 115–133 (2011)

Kim, H., Oh, E., Hahn, H.T., Lee, K.H.: Enhancement of fracture toughness of hierarchical carbon fiber composites via improved adhesion between carbon nanotubes and carbon fibers. Compos. A, **71**, 72–83 (2015)

Kim, M.T., et al.: Property enhancement of a carbon fiber/epoxy composite by using carbon nanotubes. Compos. B **42**(5), 1257–1261 (2011)

Lim, D.S., An, J.W., Lee, H.J.: Effect of carbon nanotube addition on the tribological behavior of carbon/carbon composites. Wear **52**(5–6), 512–517 (2002)

Liu, N., et al.: Effects of nano-sized and micro-sized carbon fibers on the interlaminar shear strength and tribological properties of high strength glass fabric/phenolic laminate in water environment. Compos. B **68**, 92–99 (2015)

Li, W., et al.: On improvement of mechanical and thermo-mechanical properties of glass fabric/epoxy composites by incorporating CNT–Al_2O_3 hybrids. Compos. Sci. Technol. **103**, 36–43 (2014)

Li, Y., et al.: Improvement of interlaminar mechanical properties of CFRP laminates using VGCF. Compos. A **40**(12), 2004–2012 (2009)

Lubineau, G., Rahaman, A.: A review of strategies for improving the degradation properties of laminated continuous-fiber/epoxy composites with carbon-based nanoreinforcements. Carbon **50**(7), 2377–2395 (2012)

Ma, L. et al.: Improving the interlaminar properties of polymer composites using a situ accumulation method to construct the multi-scale reinforcement of carbon nanofibers/carbon fibers. Compos. A, **72**, 65–74 (2015)

Ma, P.C., Wang, S.Q., Kim, J.K., Tang, B.Z.: In-situ amino functionalization of carbon nanotubes using ball milling. J. Nanosci. Nanotechnol. **9**(2), 749–753 (2009)

Mujika, F., et al.: Influence of the modification with MWCNT on the interlaminar properties of long carbon fiber composites. Compos. B **43**(3), 1336–1340 (2012)

Otsuka, K. et al.: Synthesis of carbon nanotubes on Ni/carbon-fiber catalysts under mild conditions. Carbon, **42**(4), 727–736 (2004)

Qiu, J., Zhang, C., Wang, B., Liang, R.: Carbon nanotube integrated multifunctional multiscale composites. Nanotechnology **18**(27), 275–708 (2007)

Rahmanian, S. et al.: Mechanical characterization of epoxy composite with multiscale reinforcements: carbon nanotubes and short carbon fibers. Mater. Des. **60**, 34–40 (2014)

Sadeghian, R., Gangireddy, S., Minaie, B., Hsiao, K.: Manufacturing carbon nanofibers toughened polyester/glass fiber composites using vacuum assisted resin transfer molding for enhancing the mode-I delamination resistance. Compos. A **37**(10), 1787–1795 (2006)

Sharma, S.P., Lakkad, S.C.: Effect of CNTs growth on carbon fibers on the tensile strength of CNTs grown carbon fiber-reinforced polymer matrix composites. Compos. A **42**(1), 8–15 (2011)

Shekar, K.C., Prasad, B.A., Prasad, N.E.: Interlaminar shear strength of multi-walled carbon nanotube and carbon fiber reinforced, epoxy—matrix hybrid composite. Procedia Mater. Sci. **6**, 1336–1343 (2014)

Shen, Z., et al.: The effects of carbon nanotubes on mechanical and thermal properties of woven glass fibre reinforced polyamide-6 nanocomposites. Compos. Sci. Technol. **69**(2), 239–244 (2009)

Siddiqui, N.A., Khan, S.U., Kim, J.K.: Experimental torsional shear properties of carbon fiber reinforced epoxy composites containing carbon nanotubes. Compos. Struct. **104**, 230–238 (2013)

Siddiqui, N.A., et al.: Manufacturing and characterization of carbon fibre/epoxy composite prepregs containing carbon nanotubes. Compos. A **42**(10), 1412–1420 (2011)

Siddiqui, N.A., et al.: Tensile strength of glass fibres with carbon nanotube–epoxy nanocomposite coating: effects of CNT morphology and dispersion state. Compos. A **41**(4), 539–548 (2010)

Song, K., et al.: Structural polymer-based carbon nanotube composite fibers: understanding the processing–structure–performance relationship. Materials **6**(6), 2543–2577 (2013)

Thostenson, E.T., et al.: Carbon nanotube/carbon fiber hybrid multiscale composites. J. Appl. Phys. **91**(9), 6034–6037 (2002)

Veedu, V.P., et al.: Multifunctional composites using reinforced laminae with carbon-nanotube forests. Nat. Mater. **5**, 457–462 (2006)

Wang, J., Zhang, X.: Poptube technology: enabling next generation multiscale and multifunctional structural composites, pp. 1774–1782. State College, Pennsylvania, USA, s.n. (2013)

Warrier, A., et al.: The effect of adding carbon nanotubes to glass/epoxy composites in the fibre sizing and/or the matrix. Compos. A **41**(4), 532–538 (2010)

Yokozeki, T., Iwahori, Y., Ishiwata, S., Enomoto, K.: Mechanical properties of CFRP laminates manufactured from unidirectional prepregs using CSCNT-dispersed epoxy. Compos. A **38**(10), 2121–2130 (2007)

Zhang, H.J., et al.: Enhanced wear properties of hybrid PTFE/cotton fabric composites filled with functionalized multi-walled carbon nanotubes. Mater. Chem. Phys. **116**(1), 183–190 (2009)

Zhang, J., Jua, S., Jiang, D., Peng, H.X.: Reducing dispersity of mechanical properties of carbon fiber/epoxy composites by introducing multi-walled carbon nanotubes. Compos. B, **54**, 371–76 (2013)

Zhou, Y., Pervin, F., Lewis, L., Jeelani, S.: Fabrication and characterization of carbon/epoxy composites mixed with multi-walled carbon nanotubes. Mater. Sci. Eng. A **475**(1–2), 157–165 (2008)

Zhu, J., et al.: Processing a glass fiber reinforced vinyl ester composite with nanotube enhancement of interlaminar shear strength. Compos. Sci. Technol. **67**(7–8), 1509–1517 (2007)

Zhu, S., et al.: Carbon nanotube growth on carbon fibers. Diam. Relat. Mater. **12**(10–11), 1825–1828 (2003)

Chapter 6
Summery

Composite materials, with carbon nanotube inclusions as the composite's reinforcing phase, are known to have many outstanding applications in the world of science and industry. Although characterization techniques of carbon nanotube based composites may have been available in other sources, less attention has been paid to gather a collection on the study of defect. Defects are inevitable in any engineered material; however, at the scale of nano, and for nanostructured materials, these defects can have tremendous impact on the properties of the nanomaterial. Hence, after knowing the defects, methods for characterizing their influence on the properties of the resulted nanocomposite both experimentally and numerically are very important to be understood. These defects cover a wide range; from randomness in distribution of fibers in the composite structure, both in terms of orientation and density, to individual CNT defects, namely atom vacancies, perturbation and doping in the structure of individual carbon nanotubes, as well as hetero-junctions and Stone-Wales defect.

It is worth highlighting that although the term "defect" seems to have a negative weight, not all defects are undesirable. In many cases, defects are not produced naturally or they are not created unintentionally during the fabrication process or purification stages, instead, they are introduced to the structure intentionally to achieve a specific goal or to receive a special reaction from the nanoengineered material for a particular application. In other words, by controlling the defects, a full range of composites with properties, very much different from those of the original matrix, can be created to serve the world of science and industry effectively.

This is not grandiloquence, only if these defects are understood and controlled, and their influences on the final material are known. Thus, in this volume, efforts have been made not only to gather a comprehensive collection about the topic and provide a brief introduction on all various aspects of CNTs and CNT-based composites, but also, to seek studies published so far that investigated the influence of the existence of some imperfections in the structure of these nanomaterials, both experimentally and computationally.

M. Rahmandoust and M.R. Ayatollahi, *Characterization of Carbon Nanotube-Based Composites Under Consideration of Defects*, Advanced Structured Materials 39,
DOI 10.1007/978-3-319-00251-4_6

In this volume, carbon nanotubes, and carbon nanotube based nanocomposites were introduced with a specific attention to the effect of defects on the mechanical properties. The applications of these nanomaterials were mentioned and some details of their fabrication and characterization methods, with a special emphasis to studies that consider the influence of the existence of some imperfections in the structure of the nanomaterials on their properties were provided. Material structure, properties and applications of carbon nanotubes and carbon nanotube based composites were also explained briefly to provide a rather comprehensive collection about the topic.

Finally, the authors strongly hope that the prepared volume would cover the necessary aspects of the topic, for those who are interested in this area of nanotechnology and we have tried our best to achieve this goal.

Erratum to: Characterization of Carbon Nanotube-Based Composites Under Consideration of Defects

M. Rahmandoust and M.R. Ayatollahi

Erratum to: M. Rahmandoust and M.R. Ayatollahi, *Characterization of Carbon Nanotube-Based Composites Under Consideration of Defects*, Advanced Structured Materials 39, DOI 10.1007/978-3-319-00251-4

The original online version for these chapters can be found at

DOI 10.1007/978-3-319-00251-4,
DOI 10.1007/978-3-319-00251-4_4,
DOI 10.1007/978-3-319-00251-4_5

M. Rahmandoust (✉)
Griffith School of Engineering, Griffith University (Gold Coast Campus), Southport, Queensland, Australia
e-mail: moones.r@gmail.com

M. Rahmandoust
Protein Research Center, Shahid Beheshti University, G.C., 1983969411 Tehran, Iran

M.R. Ayatollahi
Fatigue and Fracture Research Laboratory, Center of Excellence in Experimental Solid Mechanics and Dynamics, School of Mechanical Engineering, Iran University of Science and Technology, Tehran, Iran
e-mail: m.ayat@iust.ac.ir

M. Rahmandoust and M.R. Ayatollahi, *Characterization of Carbon Nanotube-Based Composites Under Consideration of Defects*, Advanced Structured Materials 39,
DOI 10.1007/978-3-319-00251-4_7

Preface:
The last paragraph of the preface has been modified as below.

Finally, we appreciate the moment to express our sincere gratitude to Professor Dr.-Ing. Andreas Öchsner, editor-in-chief of the Springer book series on "Advanced Structured Materials" for his helpful guidance. We would like to thank Dr. Shahin Shadlou, Dr. Mohsen Khalily, Mr. Iman Eslami Afrouz, Mr. Mehdi Mavalizadeh, Mr. Morteza Farsadi, Mr. Ali Ghavamian and Mr. Reza Barbaz for their valuable assistance. Dr. Mostafa Rahmandoust, Mrs. Azar Rezaie and Dr. Hamidreza Kazempour should also be appreciated here for their kind support. In conclusion, we are grateful to many scientists worldwide, for their novel contributions to the topic, and for devoting their lives to keep the flame of knowledge and science burning brightly and beautifully through human history. We also thank Springer publisher for recognizing the necessity of the topic and the importance of characterization of defects in nanostructured materials.

Chapter 4: Revised Figures

The legends for the figures 4.9, 4.11, 4.16, 4.24 were incomplete in chapter 4. The missed out legends have been completed and updated in the original chapter.

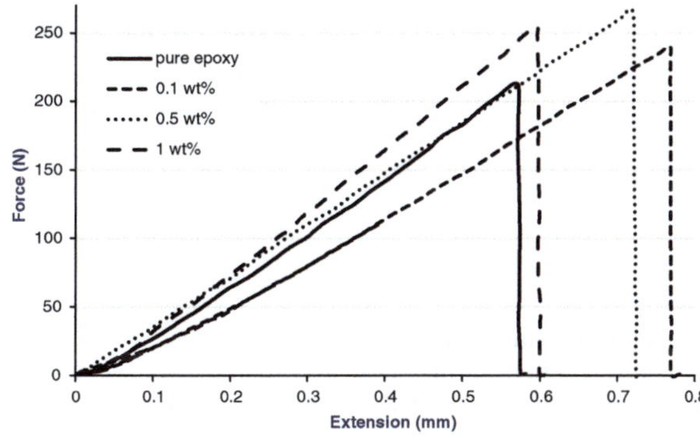

Fig. 4.9 Typical force-displacement curves obtained for pure epoxy and for nanocomposites under mode I loading conditions (Ayatollahi et al. 2011f)

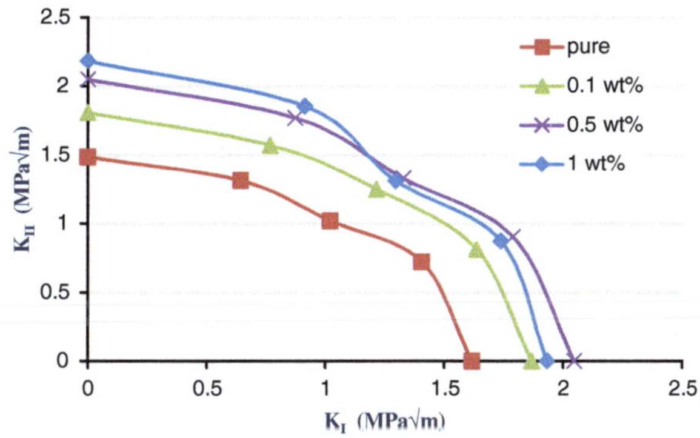

Fig. 4.11 Mixed mode I/II fracture toughness data obtained for pure epoxy and for nanocomposites (Ayatollahi et al. 2011f)

Fig. 4.16 Effects of MWCNT aspect ratio on mixed mode I/II fracture toughness data obtained for pure epoxy and for nano-composites (Ayatollahi et al. 2011g)

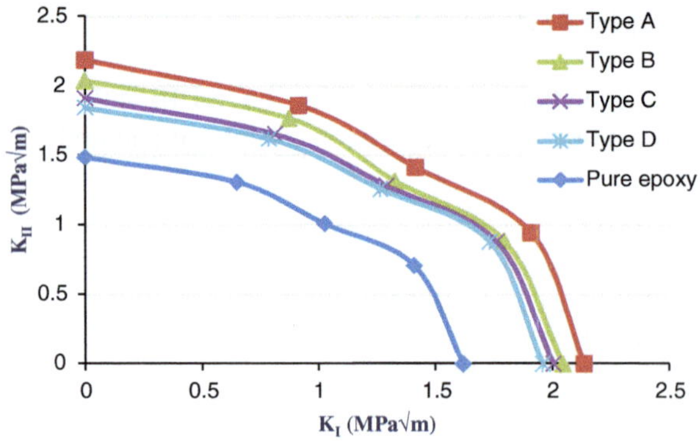

Fig. 4.24 Apparent and ploughing coefficients of friction in terms of MWCNT percent content (Ayatollahi et al. 2012a)

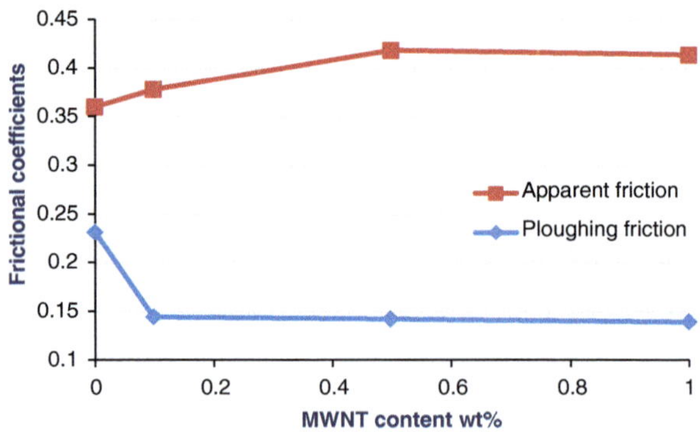

Chapter 5: Revised Figure

The legends for the figure 5.10 were incomplete in chapter 5. The missed out legends have been completed and updated in the original chapter.

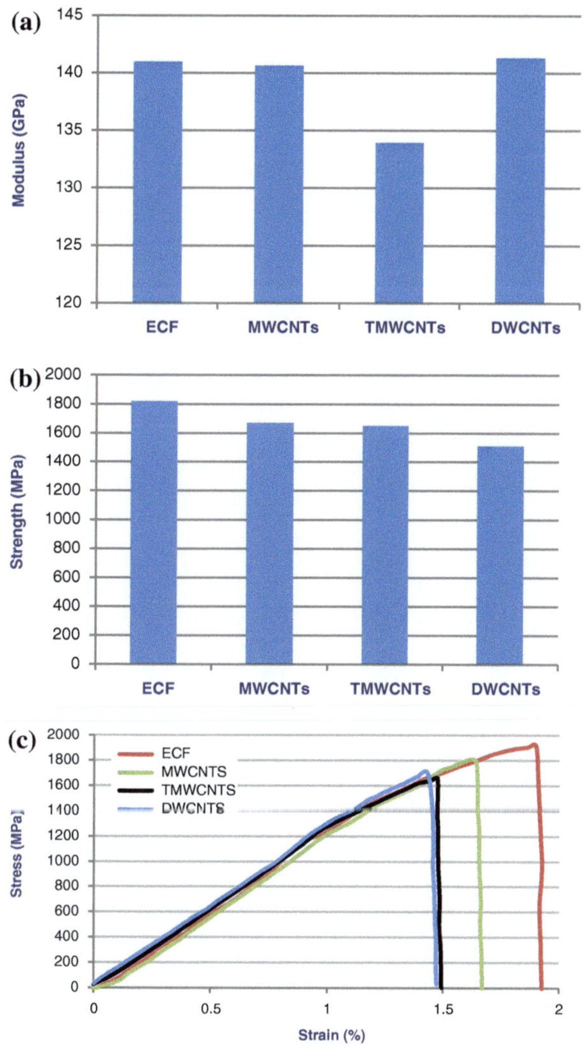

Fig. 5.10 Experimental results obtained for the tensile behavior of various UD-laminates tested in the fiber direction. **a** The tensile modulus, **b** failure strength, **c** stress versus strain curves (Godara et al. 2009)

